Tom No
630 w. 6_ st.
Leadville, CO

THE AMAZING UNIVERSE

By Herbert Friedman
Foreword by Philip Morrison

Prepared by the Special Publications Division
National Geographic Society, Washington, D. C.

The Amazing Universe

By Herbert Friedman, Ph.D., *Chief Scientist,*
E. O. Hulburt Center for Space Research,
U. S. Naval Research Laboratory

Published by
The National Geographic Society
Melvin M. Payne, *President*
Melville Bell Grosvenor, *Editor-in-Chief*
Gilbert M. Grosvenor, *Editor*
Kenneth F. Weaver, *Consulting Editor*
Philip Morrison, *Consultant, Institute*
Professor, Massachusetts Institute of Technology

Prepared by
The Special Publications Division
Robert L. Breeden, *Editor*
Donald J. Crump, *Associate Editor*
Philip B. Silcott, *Senior Editor*
Merrill Windsor, *Managing Editor*
Linda M. Bridge, Mary G. Burns, Jan Nagel
    Clarkson, Wendy W. Cortesi, Barbara
    Grazzini, Lesley B. Rogers, *Research*
Illustrations and Design
William L. Allen, *Picture Editor*
Ursula Perrin, *Art Director*
William L. Allen, Marie A. Bradby, Linda
    M. Bridge, Jan Nagel Clarkson, Wendy
    W. Cortesi, Ronald M. Fisher, Marilyn
    L. Wilbur, *Picture Legends*
John D. Garst, Jr., Virginia L. Baza, Lisa
    Biganzoli, Margaret A. Deane, Snejinka
    Stefanoff, Milda R. Stone, *Geographic Art*
Production and Printing
Robert W. Messer, *Production Manager*
George V. White, *Assistant Production Manager*
Raja D. Murshed, June L. Graham, *Production*
    Assistants
John R. Metcalfe, *Engraving and Printing*
Mary G. Burns, Jane H. Buxton, Stephanie S.
    Cooke, Suzanne J. Jacobson, Sandra Lee
    Matthews, Selina Patton, Christine A.
    Roberts, Marilyn L. Wilbur, Karen Gardner
    Wilson, Linda M. Yee, *Staff Assistants*
George I. Burneston, III, *Index*

Overleaf: Celestial stallion, the Horsehead Nebula rears in the con-
stellation Orion. Page 1: Investigating the heavens with a revolution-
ary tool—the telescope—Italian astronomer Galileo Galilei altered
man's concept of the universe. Bookbinding: Ursa Minor, the Little
Bear—or Little Dipper—holds the North Star in the tip of its tail.

Kitt Peak National Observatory rises high
above the Papago Indian Reservation in Ari-
zona. The Papagos' legendary center of the
universe, rocky Baboquivari, juts skyward
in the distance. In 1958 the tribe leased the
site to the "men with long eyes."

This rectangle indicates the part of the sky
framed in the photograph on pages 2 and
3, and the circle shows the relative size of
the full moon—both as viewed from earth.
Similar diagrams showing scale relative to the moon accom-
pany several other sky pictures in this book. Many of them
define easily visible areas, but most objects within the areas
are too faint for the unaided eye.

Foreword

I FIRST ENCOUNTERED real, professional astronomers when I was a college freshman, a good many years ago. The impression their gathering left with me was that they were enduring men and women, their patient gaze set on the eternities. The famous Princeton theorist, Henry Norris Russell, looked to me then about the age of a cliff; it came as a surprise, when I looked up the dates, to figure out he was about my own present age! Never mind the personal point of view of a youth: There was something in the notion, for the realm of the stars was slow to change. The sun evolves in billions of years, double stars rotate round each other in many decades, and even famous comets return only after many years.

Astronomy deals still in the eternities, but nowadays it cannot neglect the split-second. Dr. Herbert Friedman is an astronomer of our day; he minds the seconds as well as the centuries. His own observer's art is in rapid change, too; so that he cannot stand still on his proven technique, either. Astronomers used their eyes alone — with the aid of lens, prism, and mirror — until late Victorian years, when photography entered the science. Ever since then, the photographic plate has dominated our means of looking far away. But its unshared reign has now ended: This very book makes plain in rich photograph and crisp story how much astronomy today depends on radio reception, on novel instruments in rockets, balloons, and satellites, even on an unlikely tank of chemicals in a mine far underground, with new methods coming yearly.

It is about ten years since Herb Friedman was there on the spot at White Sands for the stopwatch time when the moon should eclipse the center of the Crab Nebula, and perhaps disclose to his rocket-borne X-ray detectors the presence of a suspected neutron star. He was there, his rocket was aloft right on time — amazing to us all in an era when the countdown was often marked by long, unexpected, frustrating periods of holding. The next opportunity meant a wait of a decade — more like the old days.

His experience stands well for today's typical investigator of the universe. We have become newly aware of fast-changing events in the heavens, like the Crab pulsar, a star flashing as fast as the ordinary TV screen, or the quasar 3C 279, which once poured out in 13 days enough energy to keep our whole Milky Way galaxy of stars shining for centuries. Fast-changing ideas, too, mark the last decade, a time when we have seen a whole new universe of unexpected objects clamor for understanding.

It is a happy, awesome, headlong time for astronomy, and this essay in clarifying word and striking image takes the reader right into the midst of it. The next novelty is sure to come soon, and our insight, we hope, not too much later.

PHILIP MORRISON, *Institute Professor*
Massachusetts Institute of Technology

Contents

Bathed in suffused light, Master Optician Norman Cole of Kitt Peak inspects the surface of a 127-inch mirror intended for an infrared telescope at Mauna Kea, Hawaii. The huge mirror later cracked before completion. Craftsmen may cut the glass for use in smaller mirrors.

The Lure of the Heavens

THE BEAUTY OF THE NIGHT SKY has been a source of fascination and pleasure for all mankind. Since time immemorial, men and women have watched and wondered at the seasonal procession of stars across the heavens, and the rhythmic wanderings of the planets among the constellations. Astronomy was the first science of ancient civilization, the avocation of Egyptian priests and Greek philosophers, medieval monks and Renaissance princes.

To the astronomer of today, probing ever deeper with mind and telescope, the universe is more than beautiful: It is amazing, violent, and endlessly mysterious. The revelations of recent research have been so dramatic and so extreme as to leave both scientists and laymen bewildered. Modern astronomy deals with the birth and death of stars; with exotic matter and fantastic energies; with near-infinities of space and time; with creation, evolution, and the ultimate destiny of the universe. As the sum of knowledge grows, the astronomer continues to seek answers to man's most profound questions: What is the grand design of the universe? How was it created? How did we get here? What are we? Are we alone?

The romance of astronomy is symbol-

Overleaf: Dawn breaks beyond the megaliths of Stonehenge on England's Salisbury Plain. Built 4,000 years ago, Stonehenge may have served early Britons as an observatory to study the sky, or as a temple to venerate it.
NATIONAL GEOGRAPHIC PHOTOGRAPHER GEORGE F. MOBLEY

ized by the graceful beauty of giant telescopes standing in serene solitude atop tall mountains. While others sleep, the lonely observer remains at his instrument, reliving the excitement of Galileo—hoping to see what no man has ever seen before. For my friend Jesse Greenstein, one of that small priesthood of scientists entrusted with the greatest telescope in the world on Palomar Mountain, California, "the observing process is an irresistible adventure ... even after 30 years. I am a telescope addict, in love with a 500-ton steel and glass monster."

Here is a science that can be high adventure: sailing halfway around the world to catch a glimpse of an eclipse from a tiny coral atoll; standing in the rocket's red glare to watch a telescope ride off into space on its fire-breathing vehicle, then listening to it report back on a radio beam from far above the murk of the lower atmosphere; crawling across the skin of Skylab to replace a film pack on the Apollo Telescope Mount; planting a camera on the surface of the moon to photograph celestial objects in ultraviolet light that can never be captured by earthbound instruments.

How does it feel to find a star that winks on and off 30 times a second, or to hear what may be the whispering echo of explosive creation 16 billion years ago, or to discover an X-ray star a hundred thousand times as bright as the sun? Ask any astronomer. The discoveries have come so fast in our generation that almost every professional has had a share.

Quasars, neutron stars, pulsars, and black holes are becoming part of the popular vocabulary. Quasars, or quasi-stellar objects, shine from the edge of the universe with power a hundred times that of our galaxy—power still unexplained. A massive star collapses to a diameter of ten miles and a density of a billion tons per cubic inch to form a neutron star. Collapse speeds up its rotation to several revolutions per second, and it becomes a pulsar, sending out flashes of radiant energy *(Continued on page 19)*

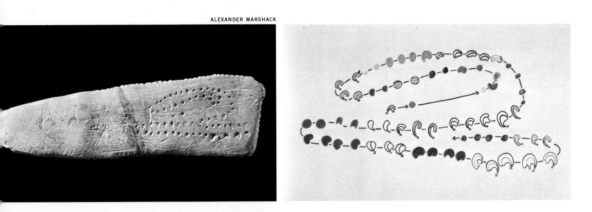

JONATHAN S. BLAIR (ABOVE); NATIONAL GEOGRAPHIC PHOTOGRAPHER JAMES P. BLAIR

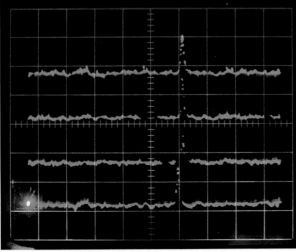

Astronomy through the ages: Engravings on a bone more than 30,000 years old suggest that it records a period of time. Archeologist Alexander Marshack, who prepared the drawing, reads the markings as a notation of the passage of 2¼ lunar months. Today, astronomers probe the universe with such facilities as those at the Mullard Radio Astronomy Observatory (above) in Cambridge, England. At left, a radio-telescope display screen records the periodic blip of a pulsar—a tiny star with an intense magnetic field. The sharp peak marks the blip; the ragged horizontal lines represent background noise.

Snowbanks pile against structures of the Pic du Midi Observatory, perched at 9,450 feet in the French Pyrenees. The thin, dry atmosphere here gives astronomers

better visibility, and the high, remote site provides freedom from the interference
of pollutants and city lights. All supplies reach lofty Pic du Midi by cable car.

13

Reaching for the Sky

Manuscript of the Maya civilization (left), dating from about A.D. *1300, records scientific observation: the relationship between the journeys of earth and Venus around the sun. Astronomers of this pre-Columbian jungle culture of Middle America translated the movements of the sun, moon, and Venus into mathematical cycles.*

Diagrams of the constellations and representations of symbolic animals circle the back of a Chinese bronze mirror (opposite). It dates from the Thang period, between A.D. *620 and 900. A poem based on the ancient Chinese conception of the universe appears on the outermost ring. It speaks of the "regularities of the Heavens, and . . . tranquillity of Earth."*

Often called the oldest scientific instrument in the world, the astrolabe measures the altitude of celestial bodies above the horizon, enabling the user to determine time and latitude. Arabic characters rim this 13th-century Moorish model.

Serpent, symbolic of eternity, supports a great tortoise, the symbol of force, in the Hindu concept of the universe (opposite). Three distinct worlds — the heavenly, the intermediate, and the infernal — rest on the backs of elephants.

MAYA MANUSCRIPT FROM "CODEX DRESDENSIS"

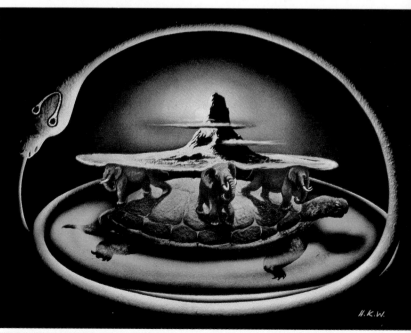

Copernicus

"In the center of all rests the sun," wrote Nicolaus Copernicus, pictured opposite with a chart of his concept of the solar system. His thesis contradicted the accepted doctrine of more than a thousand years—that the universe centered on the earth. Although Copernicus's contemporaries honored him as an astronomer, virtually all refused to accept his theory that the planets revolved around the sun. Born in 1473 in what is now Poland, Copernicus studied mathematics, law, and medicine, but spent his adult career as a church administrator. A young disciple persuaded him to publish his theories in 1543 —the year he died.

Galileo

First to use a telescope in making astronomical observations, Galileo Galilei (right, above) found proof of the Copernican theories; his discovery of four of Jupiter's moons proved that not all heavenly bodies circled the earth. On the basis of what he heard of a "magnifying tube" invented in Holland, the versatile Italian in 1609 built a telescope for himself that increased the number of visible stars from several thousand to some 50,000. He sketched much of what he saw, including several views of the moon pockmarked with mountains and craters. Church officers considered as heresy this representation of an imperfect heaven. In 1633, nine years before his death, the Inquisition forced Galileo to admit his "errors."

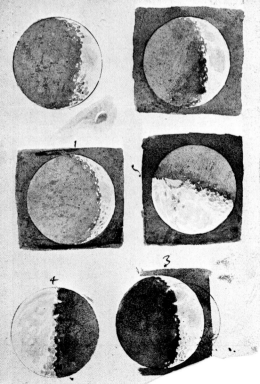

Newton

Intense scholar, creative scientist, Isaac Newton (above) studies light refracted through a prism. In a fruitful 18-month period beginning in 1665, the English genius determined the basic principles of light and color, created the mathematics of calculus, and began to comprehend gravity, the working of tides, and the motions of the planets.

Herschel

Night sky spreads above a 40-foot telescope built by William Herschel, standing outside his English home with astronomer-sister Caroline and son John, who also became a famed astronomer. The German-born Herschel discovered Uranus in 1781, determined the general shape of the Milky Way, and plotted the sun's path through space.

with startling precision. If the star is over-weight, it may collapse even further and vanish from sight — a black hole that devours both matter and light, and in the end may consume the universe.

We know of giant stars as thin as air and a hundred times as wide as the sun, of white dwarf stars harder than diamond and no bigger than the earth, of radio galaxies a trillion trillion trillion times as powerful as a home-town broadcasting station.

Astronomy is a great human endeavor, international in its ideals and performance, and shared by thousands of scientists, technicians, engineers, and craftsmen. Its instruments are the most beautifully precise, exquisitely delicate, and incredibly sensitive that the genius of man can produce. Now their capabilities are extended by computers; for the theorist, these miracle-working machines spin out the life histories of stars and galaxies, compressing billions of years of evolution into a matter of minutes. Observers continents apart synchronize their radio telescopes with atomic clocks to operate in unison from a common baseline.

Some of the greatest astronomical contributions have been made by amateurs, and many of the grossest misconceptions have come from experts. Certainly there has been no single path to achievement. In 1917 a former mule driver, Milton Humason, started work at Mount Wilson as a janitor, and soon began to assist with observations. Eventually he became one of the most skilled of astronomers at the highly specialized tasks of investigating expansion of the universe.

Patrons of astronomy have come from all walks of life. Industrialist Andrew Carnegie was the generous benefactor of Mount Wilson Observatory. The wealth of James Lick, a piano-maker who amassed a fortune from real estate investments, created the Lick Observatory, the most enduring memorial he could conceive, although he had no connection with astronomy during his lifetime. His body is entombed in the pedestal of the observatory's original telescope.

IN THE SECOND CENTURY A.D. the Greek astronomer Ptolemy — Claudius Ptolemaeus of Alexandria — completed his treatise the *Almagest*, which described the regular movements of the planets against the fixed stars. In Ptolemy's universe the heavens revolved about a flat, motionless earth. All the stars were fixed to a hollow sphere which made one revolution every 24 hours on an axis passing through the earth. Sun and moon were attached to smaller transparent spheres, each centered on earth and turning at a different speed. Each of the five planets then known rode on its special sphere, fastened to the rim of a wheel whose axis was attached to the large transparent sphere. Even this arrangement was not precise enough, and Ptolemy added wheels within wheels.

For 14 centuries this complicated explanation of planets moving in cycles and epicycles about a fixed earth was generally accepted as authoritative. There was little further advance of significance in astronomy until 1543, when the churchman and scholar Nicolaus Copernicus published *On the Revolution of the Celestial Orbs*.

With this remarkable work Copernicus denied prevailing dogma and dethroned the earth from its position at the center of the universe. But his revolutionary concept that the earth and planets revolve about the sun was not readily accepted by the medieval intellect. Martin Luther was scornful, pointing out that Joshua had commanded the sun, not the earth, to stand still.

Early in the 17th century, the German astronomer Johannes Kepler reinforced the Copernican revolution with his determination that the orbital paths of the planets were elliptical. Then, in 1609, the myopic views of the Middle Ages were challenged dramatically when Galileo Galilei pointed his crude telescope toward the sky and discovered undreamed-of wonders. Mountains rose on the moon, and craters pitted its surface. Bright Venus waxed and waned with phases resembling those of the moon, proving that it shone by *(Continued on page 25)*

A Telescope Mirror Takes Shape

Twenty-five tons of molten glass, hot enough to singe a backdrop of asbestos, glows just after pouring at the Owens-Illinois Glass Works in Toledo, Ohio. In a process requiring years, this glass became a precision scientific instrument—a 158-inch mirror for a telescope at the Cerro Tololo Inter-American Observatory in Chile. After the mirror cooled, technicians rough-ground and shaped it, then transported it to the Kitt Peak National Observatory Optical Shop in Tucson, Arizona. There, 2½ years of grinding and polishing (lower left) gave the mirror —now reduced to 17 tons—a slightly concave surface. At center, a technician lowers a Hartmann screen over the mirror; it deflects light onto the surface to test reflectivity. Finally, scientists flood the mirror with a laser's ruby glow to seek microscopic flaws. Irregularities as slight as five millionths of an inch show up as dark blotches.

HERRAL W. LONG (ABOVE); N.G.S. PHOTOGRAPHER JAMES P. BLAIR (LOWER LEFT); JOHN LUTNES, KITT PEAK NATIONAL OBSERVATORY (CENTER); N.G.S. PHOTOGRAPHER JAMES P. BLAIR AND NORMAN C. COLE, KITT PEAK NATIONAL OBSERVATORY

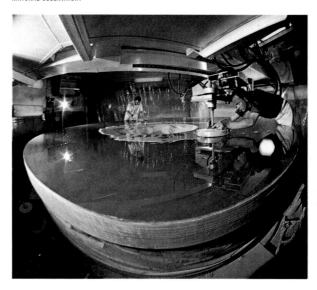

KITT PEAK NATIONAL OBSERVATORY

Journey to a Mountaintop

Topographic computer maps of the Cerro Tololo mirror (above) show progressively smoother surfaces achieved by grinding and polishing; the lower right map shows few defects in the completed mirror. Carefully packed and hauled by truck (opposite), the mirror passes through La Serena, Chile, during the trip to 7,200-foot Cerro Tololo (opposite, above), some 250 miles north of Santiago. Scientists selected the site for its clear and stable air, and to give astronomers a large observatory in the Southern Hemi-sphere; about a fourth of the sky—including the region of our nearest galactic neighbors, the Magellanic Clouds—cannot be studied from the Northern Hemisphere. The Association of Universities for Research in Astronomy (AURA) operates the facility; astronomers from many countries share it. The first photograph made with the new telescope (far right) captured 47 Tucanae, brightest globular star cluster known. The "white sky" image shows stars as black points, the way they appear on the photographic plate.

DAVID L. MOORE, KITT PEAK NATIONAL OBSERVATORY (ABOVE); CERRO TOLOLO INTER-AMERICAN OBSERVATORY (LOWER LEFT); DR. VICTOR BLANCO, CERRO TOLOLO INTER-AMERICAN OBSERVATORY

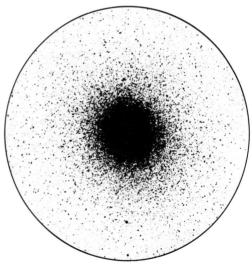

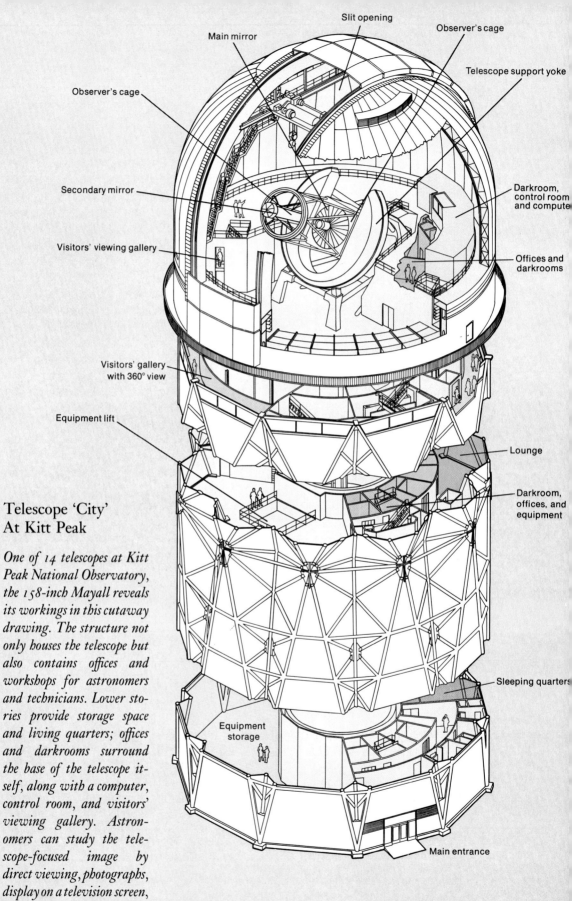

Slit opening

Observer's cage

Main mirror

Observer's cage

Telescope support yoke

Secondary mirror

Darkroom, control room and computer

Visitors' viewing gallery

Offices and darkrooms

Visitors' gallery with 360° view

Lounge

Equipment lift

Darkroom, offices, and equipment

Sleeping quarters

Telescope 'City' At Kitt Peak

One of 14 telescopes at Kitt Peak National Observatory, the 158-inch Mayall reveals its workings in this cutaway drawing. The structure not only houses the telescope but also contains offices and workshops for astronomers and technicians. Lower stories provide storage space and living quarters; offices and darkrooms surround the base of the telescope itself, along with a computer, control room, and visitors' viewing gallery. Astronomers can study the telescope-focused image by direct viewing, photographs, display on a television screen, or special instrumentation.

Equipment storage

Main entrance

DRAWING BY SNEJINKA STEFANOFF, FROM PLANS
BY SKIDMORE, OWINGS, AND MERRILL

reflected sunlight and that it was indeed a satellite of the sun. The sun itself, the most unblemished object of Aristotelian philosophy, was blotched with dark spots—and rotated. And four moons went around and around Jupiter as the planet moved steadily across the heavens. These discoveries dealt a crushing blow to Ptolemaic theory.

In England, toward the end of the century, Isaac Newton derived and explained the precise shapes of planetary orbits as being a natural consequence of the force of gravitation: The motions of planets, he said, are governed by the same law as the fall of an apple from tree to ground. For the next 200 years, mathematicians elaborated on Newton's work and compiled a remarkably accurate description of the motions of the celestial bodies.

Yet at speeds approaching that of light, even Newtonian physics breaks down. In 1905, in Switzerland, Albert Einstein produced his Special Theory of Relativity, which applied at all velocities. By 1916 he had expanded this into the General Theory of Relativity—a new concept of gravitation widely acclaimed as the most esthetically satisfying intellectual achievement of the mind of man. It has met all the subtle observational tests failed by Newtonian theory. Nevertheless, the neutron stars, pulsars, and black holes of contemporary astronomy will pose newer and even more crucial tests.

HISTORICALLY, astronomy and physics have been closely related, and in the present century the specialty of astrophysics has become a major discipline. To understand the cosmos, we must understand the world of the very small—the elementary particles and their combinations in nuclei, atoms, and molecules. For it is the behavior of these minute components, as revealed through spectroscopic analysis of their radiation, that has taught us most of what we know about the nature and composition of the stars.

Until the end of the 19th century, 2,300 years after the Greek philosopher Democritus first introduced the concept of the atom, it was still considered to be the ultimate, indivisible building block of matter.

The first fundamental subdivision of the atom to be discovered was the electron, a particle bearing a single negative electrical charge. Free electrons carry electric current in metals and in radio tubes. In a television picture tube, a beam of electrons creates a bright dot of fluorescent light and draws an image as the beam rapidly scans a series of horizontal lines. Neon signs carry luminous discharges produced by electrons rushing through low-pressure gas and transferring energy in collisions with the atoms.

After scientists identified the electron, they unlocked the secret of the proton—the nucleus of the hydrogen atom and a part of the atomic nucleus of every other element. The proton carries a single positive electrical charge, but its mass is nearly 2,000 times as great as that of the electron.

Then, in the early 1930's, the neutron was discovered. Its mass is just slightly greater than that of the proton, but it carries no electrical charge.

Since then, well over a hundred new particles have been produced in the laboratory or found in cosmic rays. The words "elementary particle" hardly have meaning any more. But for most astrophysical problems, it is sufficient to treat ordinary material as collections of electrons, protons, and neutrons. Under normal conditions, electrons are bound up in atoms; the positively charged protons in the nucleus attract an equal number of negatively charged electrons, and the atom itself is electrically neutral. Under certain circumstances, however, atoms lose or gain electrons and become "ionized"—positively or negatively charged.

In a sense an atom resembles a submicroscopic planetary system with the electrons orbiting the nucleus at relatively great distances. If we could magnify the nucleus a trillion times, it would be the size of a

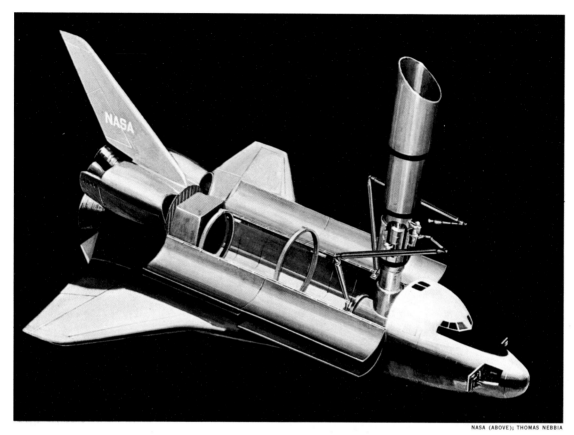

NASA (ABOVE); THOMAS NEBBIA

Shuttle Astronomy

Delta-winged Space Shuttle may one day carry a powerful telescope into near space, a program long advocated by Lyman Spitzer, Jr., (right) of Princeton University. In the painting above, a 94-inch telescope has emerged from the shuttle's bay; an astronaut makes final adjustments. Such a telescope, far above the blurring effect of earth's atmosphere, could study distant stars with a clarity only dreamed of by earthbound astronomers. The reusable shuttle, scheduled to fly in the early 1980's, will carry crews to operate a space laboratory and launch, recover, and repair satellites. It will house scientific experiments and haul people and cargo. After a mission, it will return to earth and land like an airliner.

pinhead. At this same scale each electron would revolve in an orbit that would just fit inside a large room. In the microcosm of the atom, the space inside the electron orbit is as empty as the space between the planets and the sun.

TO THE UNAIDED EYE several thousand stars are visible in the night sky. All of these belong to our galaxy, the Milky Way. At the turn of the century, man's concept of the universe was still very limited; nothing was known to exist beyond this galaxy of stars, and astronomers believed intuitively in their unchanging nature.

Now, only three-quarters of a century later, we know that the pattern of the universe is fantastically greater and more complex. On cosmic time scales, stars and even galaxies are ephemeral—glowing, flashing, fading, and eventually disappearing into inky blackness. Continuous transformation is the natural order, even though the firmament appears unchanging in the minuscule time frame of human life.

The energy of the stars is not inexhaustible. It derives from thermonuclear fusion—the union of atomic nuclei under intense heat and pressure—and limits the life of stars like the sun to about ten billion years. Our sun is only one of about 200 billion stars in the Milky Way. Today's great telescopes search out billions of other galaxies, some of them billions of trillions of miles distant. The galaxies are rushing away from each other like fragments from a titanic explosion. As we look deep into space we look backward in time almost to creation, when the light from the most distant galaxies started its journey toward earth.

Cosmology, or the study of the age, size, and mass of the universe, has proposed three principal theories in recent years: "explosive creation," or the "big bang," as it was dubbed by George Gamow; the "repeated big bang" or "oscillating universe," also given prominence by Gamow's discussions; and the "steady-state universe" of Herman Bondi, Tom Gold, and Fred Hoyle.

The scenario of the big bang assumes that all the material of the universe was originally bound in one super-hot, super-dense mass. Its temperature was higher than a trillion degrees, and all matter was homogenized into hydrogen and helium. The explosion of this fireball kicked out matter with enormous velocity in all directions. A billion years later, the clouds of matter began to condense into galaxies of stars. Our Milky Way, at the age of 15 billion years, is still evolving new stars.

Everything in the solar system—including our own bodies—has been reprocessed from material left behind by earlier stellar explosions. The sun was born about five billion years ago. Just as planets formed about the sun, innumerable other planets probably surround billions of other stars in the Milky Way. With the evolution of a planetary geological structure and the development of a stable planetary atmosphere, life becomes possible; similar evolution may have occurred in millions of other planetary systems. We have no basis to presume that we are the first and only specimens of intelligent life, or the latest, or by any means the most advanced.

If the universe has been expanding with escape velocity from a primordial explosion—that is, with sufficient speed to overcome indefinitely the counter-effects of gravitational force—all the lights will eventually go out, leaving the dead galaxies strewn in an infinite graveyard of space. But if the force of expansion is weaker than the gravitational pull of all the galaxies acting on each other, the expansion should eventually come to a halt, and all the universe should fall back on itself. This is the plot of the repeated-big-bang theory: Perhaps the return to the fireball condition will be followed by a new explosion, and the universe will alternate between expansion and contraction.

The steady-state cosmology suggests a universe that has no beginning and no end. It retains its uniformity by continuous creation which may take place atom by atom or a galaxy at a time. As newly created material is fed into the universe, it will force expansion. If the rate of adding new matter increases, expansion must speed up; if the rate decreases, the expansion slows. If the gain is properly balanced, a steady average density can be maintained. The universe then continues to grow, but always looks the same.

These different cosmologies have different esthetic appeals. The big-bang universe requires a unique switching-on, which may satisfy religious instincts of an existence external to the universe. The oscillating universe implies a certain rhythm of repetition: All that develops in one 80-billion-year cycle is wiped out by the end of the cycle, and then the same performance is repeated. The steady-state universe sets aside all questions of beginning and ending.

At present, the evidence favors the big bang. But no matter how far we push back the veil of ignorance to reveal the existing universe, we shall always be stumped by the basic issue: What lies behind creation?

THE NEW ASTRONOMY matches farsighted telescopes with far-reaching theoretical concepts to discover and interpret a continuous flood of startling information about our universe. Optical telescopes can enhance the sensitivity of the human eye nearly a million times. Radio telescopes in tandem produce images a thousand times sharper than those of the greatest optical instruments. Rockets, balloons, and satellites carry infrared, ultraviolet, X-ray, and gamma-ray telescopes beyond the atmospheric veil that blankets the earth. No message from the depths of the cosmos on any frequency of the entire radiation spectrum is now beyond our capability for reception.

The adventure into space with rockets and satellites is a fulfillment of astronomers' dreams, and holds promise of wonderful opportunities to come. Ideas of space-ship observatories have intrigued the human imagination ever since Galileo. In 1726,

Jonathan Swift wrote about the flying island of Laputa from which observations were made of the moons of Mars. Laputa was essentially an orbiting astronomical space station. The Russian dreamer of rockets and space travel, Konstantin Tsiolkovsky, wrote nearly a century ago: "To step out onto the soil of asteroids, to lift with your hand a stone on the moon. . . . to land on [Martian] satellites and even on the surface of Mars — what could be more extravagant! However, it is only with the advent of reactive vehicles that a new and great era in astronomy will begin, the epoch of a careful study of the sky. . . ."

More than 50 years ago the German rocket pioneer, Herman Oberth, whose childhood fantasies had been inspired by the tales of Jules Verne, published an influential book called *Rocket into Interplanetary Space*. He pointed out all the major advantages of an astronomical telescope in orbit.

American astronomer Lyman Spitzer, Jr., long a courageous advocate of space observatories, has seen his ideas come to fruition in the NASA-Princeton University Orbiting Astronomical Observatory named *Copernicus*. And he is deeply involved in planning for the Large Space Telescope, scheduled to be launched on a Space Shuttle in the early 1980's. It now seems clear that much of astronomy's exciting future lies in observation from space.

In the following chapters, we shall see how physics and astronomy have made the world over in this century. The new picture of the universe transcends all previous imagination, but it may be only prologue to more astounding revelations yet to come. Scientific speculation is rife. A scent of impending Copernican revolution is in the air.

Fogbank rolls below Fremont Peak near Monterey, California, as members of the San Francisco Sidewalk Astronomers view the planet Venus. They use the 24-inch Delphinium — world's largest portable telescope.

To the Edge of the Universe

To begin a description of our amazing universe, we must try to appreciate its vast scale. "Try" is the right word, for astronomical distances are so great that it is extremely difficult—with our everyday, human concepts of time and space—to sense them at all. Because of the immensities involved, astronomers find a convenient measure in the speed of light: 186,282 miles a second.

A ray of light travels from moon to earth in slightly more than a second, so the earth-moon distance—240,000 miles—can be expressed as about 1.3 light-seconds. The 93 million miles from sun to earth is the equivalent of 8.3 light-minutes. A light-year, the astronomical measure commonly used, is about *six trillion* miles.

Since today's rockets never achieve more than a small fraction of the speed of light, space travel requires about three days to the moon, two years to Jupiter, and 15 years to Pluto. To reach the nearest star, Alpha Centauri, at a distance of 4.3 light-years, would take nearly 100,000 years!

Stand out of doors in the country on a clear night and look up: Without a magnifying lens you can see several thousand stars. A pair of binoculars will bring into view per-

Overleaf: Illuminated by a brilliant star cluster, a nebula forms a celestial abstract in the constellation Serpens. Spectacular nebulae —enormous clouds of gas and dust—often screen from view the vast universe beyond.

KITT PEAK NATIONAL OBSERVATORY

haps 50,000. If you have a two-inch telescope, the number leaps to several hundred thousand. The current estimate is that our galaxy, the Milky Way, contains about 200 billion stars, many of them in clusters of hundreds of thousands.

The galaxy itself takes the form of a disk with a bulge at the center; the diameter of the disk is about 100,000 light-years (or 600 quadrillion miles), its central thickness is about 10,000 light-years, and it tapers rapidly to about a thousand light-years near our sun. The galaxy appears to us to arch across the sky in a powdery band of light.

The Milky Way is part of a local family or cluster of about 20 galaxies, including Andromeda and the two Magellanic Clouds, that lie within a range of about three million light-years. More than ten billion galaxies fill the observable universe, whose radius is estimated at some 16 billion light-years. No matter where we look, we see more galaxies. If there is a limit to the universe, the Milky Way must be far from its edge; but whether near the center or far from it, we have no way of knowing.

The universe extends so far that we need simplified comparisons to grasp even its relative dimensions. Suppose the scale were reduced a trillion times so that the sun became the size of a pinhead; the entire solar system then would fit inside a large living room. Twenty-six miles away would lie the nearest star. The Milky Way would be a disk about 60,000 miles thick at the center and 600,000 miles in diameter, dotted with 200 billion sparkling pinheads.

Even on such a scale, the universe beyond this galaxy strains our comprehension. Reduce the Milky Way again, from 600,000 miles across to the size of a phonograph record. Now the nearest spiral galaxy, Andromeda, would be a similar disk spinning in space some 20 feet away. Billions of miniaturized galaxies—disk-shaped, elliptical, and irregular—would stretch out to the edge of the observable universe, nearly 30 miles distant.

Midway through the 19th century, astronomy had largely settled into a routine of cataloguing thousands upon thousands of stars. Their positions and relative motions were determined with considerable precision, but all stars were treated simply as pointlike objects. Why stars shone, and what they were made of, seemed destined to remain forever unanswered. Man must reconcile himself, declared the French philosopher Auguste Comte, to eternal ignorance of the composition of the stars.

How wrong he was! Today, by analyzing the color spectrum of starlight, we can tell what comprises a star as accurately as if part of it had been brought to earth. With a special instrument called a spectroscope, the astronomer passes light through a narrow slit and resolves it into its various colors by means of a glass prism or a finely ruled diffraction grating on a metal-coated mirror surface. Each separate wavelength of light appears in the spectrum as a distinctive color shade; and each chemical element transmits a characteristic pattern of spectral lines — its individual "fingerprint."

The history of spectroscopy begins with Isaac Newton. In the year 1666, he later recorded, "I procured a triangular glass prism, to try therewith the celebrated phenomena of colours. And for that purpose having darkened my chamber, and made a small hole in my window shuts, to let in a convenient quantity of the sun's light, I placed my prism at [its] entrance, that it might be thereby refracted to the opposite wall. It was...very pleasing...to view the vivid and intense colours...." Newton also found that he could reproduce white light by passing all the rainbow colors through a second prism.

From that simple start, spectroscopic analysis has developed to a stage where the nature of a star's spectral lines lets the astrophysicist deduce temperature, pressure, density, and composition; the strength of gravity, electric force, and magnetic force; the degree of (Continued on page 39)

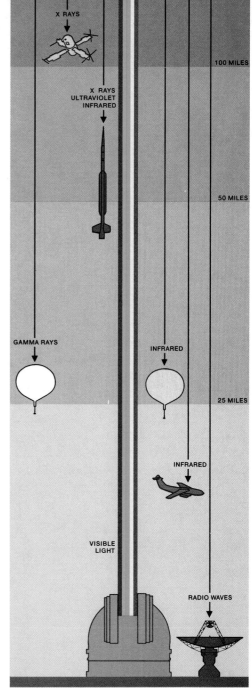

GEOGRAPHIC ART

Electromagnetic spectrum helps reveal secrets of the universe. Measured in wavelengths, radiation (above, left to right) ranges from short gamma rays to long radio waves. Earth's atmosphere blocks most wavelengths (arrows indicate penetration), and astronomers long observed only the narrow band that reaches earth as visible light; in the 1930's scientists also detected radio waves from space. Now airplanes, balloons, rockets, and satellites carry instruments aloft for unobscured observation of celestial radiation.

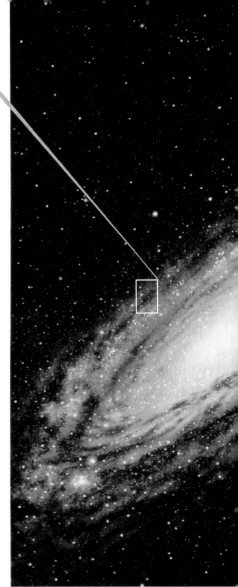

The Earth in Perspective

Awesome ... astounding ... infinite ... words falter in describing the immensity of our universe. The vastness of earth's surroundings unfolds in this sequence of paintings: With each successive look, the observer takes a giant step back in space. A small rectangle locates each picture, reduced to scale, in the next view. Above, the planets—their size exaggerated in relation to neighboring stars—appear on an imaginary line approximating their orbital planes. Earth spins to the left of our radiant sun—93 million miles away—while Mercury, Venus, Mars, and Jupiter extend right. This area, when placed in its niche near the Milky Way's rim in the next picture, shrinks drastically: Our sun decreases to a speck of gold dust among the galaxy's 200 billion stars; the planets become invisible. Yet even the brilliant Milky Way, seen in the following view in its local cluster of galaxies—each with billions of stars— diminishes to a small glow. In the final painting this cluster, placed in a far wider setting that still represents a tiny fraction of the universe, sparkles like a sprinkling of diamond chips. Poet Archibald MacLeish put our earth into perspective: "a small ... planet ... of a minor star ... at the edge of an inconsiderable galaxy in the immeasurable distances of space."

PAINTINGS BY DAVIS MELTZER

Series of lightning bolts, captured by a time exposure, illumines Kitt Peak. One jagged fo
frames the domed, 19-story structure housing the 158-inch Mayall Telescope—one

GARY LADD

be three largest optical telescopes in the world. The Mayall, used as a giant camera,
an record objects so distant that their light has taken ten billion years to reach us.

37

Moving our sun trillions of miles seems a Herculean task, but Harlow Shapley (above) accomplished it in 1918 with an inspired assumption. Until then, astronomers guessed at the size of the Milky Way, usually placing the sun at its center. But Cepheid variables — stars with fluctuating light — had provided a clue: They brighten and dim over precise time periods, and their periods and brightness yielded a yardstick of astronomical measurement. Using Cepheids as distance indicators, Shapley plotted the positions of masses of stars called globular clusters. His theory that their distribution indicated the galaxy's center — putting our sun out toward the edge — proved correct, but his estimate of the galaxy's huge dimensions shocked his colleagues. In an edge-on drawing of the Milky Way (below), large dots represent globular clusters; a dark line indicates the veil of obscuring dust along the galactic plane. Our sun, at a distance of 30,000 light-years, takes 200 million years to complete one orbit of the galaxy's center.

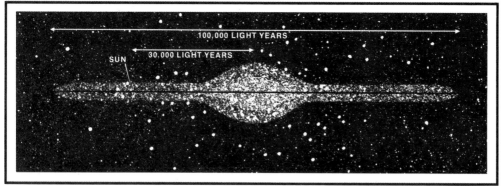

turbulence, and convective heat movements in a specific region of the star.

At the time of Newton's experiments, it was believed that the sun's light contained its heat. About 1800 William Herschel, while working with colored filters on his telescope, found that he sometimes felt heat even when most of the sun's light was blocked out. Placing a thermometer in different parts of the spectrum dispersed by a prism, he detected little temperature increase in most of the color range; but as he moved the thermometer into the red and then beyond, where no light could be seen, the temperature rose rapidly. It was the first recognition of infrared rays.

In 1801 Johann Wilhelm Ritter, experimenting in Germany with light's effect on silver salts, discovered invisible ultraviolet rays when silver chloride placed beyond the violet end of the spectrum turned black.

The Scot James Clerk Maxwell developed the modern theory of light as an electromagnetic wave in the mid-19th century. His theory clearly predicted it should be possible to generate man-made electromagnetic waves; and by 1888, nine years after Maxwell's death, Heinrich Hertz succeeded in producing radio waves.

In 1895 German physicist Wilhelm Roentgen detected invisible radiation that could penetrate flesh and cast shadows of bones on a fluorescent screen. Unable to explain the phenomenon, he coined the term "X rays." And in 1896, Antoine Henri Becquerel of France discovered radioactivity when his photographic plates, wrapped in heavy lightproof paper, were fogged in the dark by uranium he had placed nearby. Four years later his countryman Paul Villard detected ultrashort gamma rays.

Thus we have learned that we live in a dual universe—part of it we see, and part we don't. What the eye detects is only a minute range of wavelengths and frequencies in nature's vastly broader spectrum of electromagnetic radiations. By analogy with pitch on the musical scale, the frequencies of visible colors span not quite one octave; the full electromagnetic spectrum spreads over about 84 octaves.

Wavelengths range from more than a thousand miles to far, far less than a trillionth of an inch. Because this range is so great, it is common to measure spectral regions in units of kilometers, meters, centimeters, and millimeters for radio waves; micrometers (microns) and angstroms for infrared, visible, ultraviolet, X rays, and gamma rays. One micron is a millionth of a meter, and one angstrom is a ten-billionth of a meter. The wavelength of visible green light, for example, is about 5,400 angstroms.

THE AIR ABOVE US makes life livable and comfortable, but it blinds the astronomer to all but a very limited view of the universe. Besides blocking out most rays of the electromagnetic spectrum headed our way from space, the atmosphere is turbulent and causes the twinkling appearance of stars. In an effort to escape atmospheric interference, astronomers launched telescopes to balloon altitudes, but even at those heights the atmosphere blocked most of the ultraviolet rays, X rays, and gamma rays. It seemed particularly frustrating that radiation that had traveled billions of light-years could be lost in the last thousandth of a light-second, barely a hundred miles from completing its journey to earth.

The age of space exploration has brought a solution. Rockets and satellites can now place telescopes and photometers above the veil of the atmosphere, and the universe has been observed from end to end of the electromagnetic spectrum.

The attempts to carry instruments and men above the atmosphere have, unfortunately, been marked by a good share of human tragedy. One of the most ironic was the flight on May 4, 1961, of two naval balloonists, Comdr. Malcolm Ross and Lt. Comdr. Victor Prather, to a record altitude of 113,740 feet. After the balloon had been brought down successfully, Commander

Prather slipped from the helicopter sling and drowned.

Contrasting success and failure with space vehicles has sometimes influenced the future professional paths of individual scientists. Jesse Greenstein was a young astronomer at Yerkes Observatory, Wisconsin, when the first V-2 rockets were brought to the United States from Germany. With great enthusiasm he helped prepare an ultra-violet spectrograph—a spectroscope fitted with photographic film—for solar studies. Then, in flight, the shutter failed to open, and the experiment was a total loss; in disgust Jesse abandoned this unreliable new field. But rocket astronomy's loss was classical astronomy's gain; Greenstein's research at Palomar has made a huge contribution to our knowledge of stellar evolution.

By contrast, my own first rocket astronomy experiment in 1949 was a shining example of beginner's luck: We got room aboard only because of the late dropout of another experiment, and our hastily assembled package worked perfectly.

For my next try, a year later, I built an elaborate instrument assembly. As the firing button was pushed, I stood at the blockhouse window at White Sands, New Mexico, and saw a great burst of flame and a billowing cloud of black smoke. Paralyzed, I watched as the huge rocket began to crumple and slowly toppled toward the blockhouse. When the wreckage had settled to the ground, I suddenly realized that I was the only one on my feet—everyone else had hit the deck or scrambled for cover.

FOUR HUNDRED YEARS after Copernicus removed the earth from the central position in the solar system and put the planets in orbit about the sun, a Missourian named Harlow Shapley accomplished a parallel feat: He disproved the popular concept that the sun was at the center of the Milky Way.

Interest in the distribution of stars had begun to grow by the mid-18th century. In 1755, philosopher Immanuel Kant recognized the flattened form of the galaxy—that it stretched out over a great length but was relatively thin. He also made the remarkable guess that such cloudy patches of visible nebulosity as Andromeda were themselves separate entities, or "universes," comparable to the Milky Way.

About 1780 William Herschel, a former German military bandsman whose life in England had been devoted to performing and composing music, turned to studies of the sky. Positional astronomy at the time seemed to be working out to perfection. Commenting on Newtonian mechanics and the precision with which the motions of planets and stars could be calculated, Frederick the Great of Prussia reportedly remarked that everything of importance in science had been discovered.

But Herschel was not particularly concerned with positional accuracy; his interest was exploratory in the broadest sense. For that purpose he constructed a 20-foot telescope of 19-inch aperture, and undertook a systematic counting of the stars. From this he tried to gauge the distance to the edge of the Milky Way. In the direction where the numbers were relatively few, he concluded we were close to the edge; where the counts were greatest, he assumed he was looking down the longest dimension to the rim. Herschel estimated that the Milky Way contained many millions of stars—not a bad guess, given the period and circumstances of his work.

The versatile Herschel characteristically applied his talents in a big way. The festival orchestras and choruses he conducted at Bath were among the largest assembled in England. When it came to telescopes, he was as fascinated by the challenge of building the largest instrument as by its eventual use. With a grant of £4,000 from King George III, he designed and constructed a telescope of 40-foot focal length and 48-inch aperture—comparable in size to the 48-inch Schmidt instrument that in the 1950's gave us what is still our best working map of the heavens, the National Geographic Society-Palomar Observatory Sky Survey.

Herschel was an extraordinarily prolific worker. He presented his first catalogue of a thousand nebulae and star clusters to the Royal Society in 1786, and followed it with two extensive additions. John Herschel continued his father's work, and in 1864 completed *The General Catalogue of Nebulae*, with 5,079 objects, of which 4,630 were discovered by father or son.

After William Herschel's death in 1822, little progress was made in establishing the true position of the sun in our galaxy until Shapley, a farm boy with a doctorate from Princeton, arrived at Mount Wilson in 1914. Shapley had grown up at the edge of the Ozark country, and went to work at 15 as a crime reporter on the Chanute, Kansas, *Daily Sun*. Two years later he was refused admission to high school in Carthage, Missouri, as not qualified. One of his early enthusiasms was poetry—"poetry you could recite while milking a cow and keep the rhythm going." An essay on Elizabethan verse helped gain him admission to the University of Missouri, where he discovered astronomy.

The constellation Cepheus the King lies opposite the Big Dipper on the other side of Polaris, the North Star. It does not appear especially interesting except for the marked variations in the light of one of its members, Delta Cephei. Over a period of five days and nine hours, with clocklike regularity, that star passes from bright to relatively faint and back to bright again with a contrast so pronounced that it is evident to the unaided eye. Its light variations are caused by fluctuations in size: increasing in brilliance as the star swells, dimming as it shrinks. Delta Cephei was first observed in 1784; and in due course astronomers found many more such stars, some with longer periods, some shorter, and labeled them Cepheid variables.

In 1905 Henrietta S. Leavitt, an astron-omer and head of photographic stellar photometry at Harvard College Observatory, received a collection of photographs from Harvard's southern observatory in Peru—repeated nightly exposures of the Magellanic Clouds made at her request. From them she was able to detect numerous variable stars. In the Small Magellanic Cloud she found periods ranging from as short as 1½ days to as long as 127. The longer the period, the brighter was the star.

The relationship of apparent brightness to distance was well understood. Watch a man carry a lamp away from you, and its apparent brightness will decrease fourfold with each doubling of distance. Thus, if he bears a standard lamp, you can calculate his distance as far as the lamp can be detected, by measuring its apparent brightness.

In the case of Miss Leavitt's variable stars, since all were in the Small Magellanic Cloud it could be assumed that all were about the same distance from earth. It followed that their fluctuation periods could be taken as measures of absolute brightness. Then, by independently establishing the distance to one nearby Cepheid in the Milky Way, and relating this distance to the star's apparent brightness and fluctuation period, astronomers had a standard candle. Miss Leavitt's work had resulted in a new astronomical yardstick.

With the Cepheid variables as his standards, Harlow Shapley began to establish the true dimensions of the Milky Way. Scattered about the galaxy are great concentrations of hundreds of thousands of stars known as globular clusters. In the first quarter of the 20th century some hundred globular clusters were known, and Shapley devoted himself to studying their variable stars. He derived distances for 69 clusters, and found that they were distributed about a common point which, he reasoned, must be the center of the Milky Way. The sun was near the edge of this system—and fully two-thirds of the distance to the outer edge of the galaxy.

Probing night skies during the early 1920's, Edwin Hubble (left) photographed two relatively nearby spiral nebulae with the new 100-inch telescope on Mount Wilson. He measured the distances of these objects and established them as star systems — not clouds of gas — outside our galaxy. Since many astronomers believed the cosmos consisted only of the Milky Way, Hubble's work yielded a dramatic result: proof of a much vaster universe. Continuing to survey the heavens, Hubble devised a classification of galaxies still in use. Below, the Hercules cluster of spiral and elliptical galaxies shines far beyond a scattering of Milky Way stars.

COURTESY OF MRS. EDWIN HUBBLE (ABOVE); HALE OBSERVATORIES

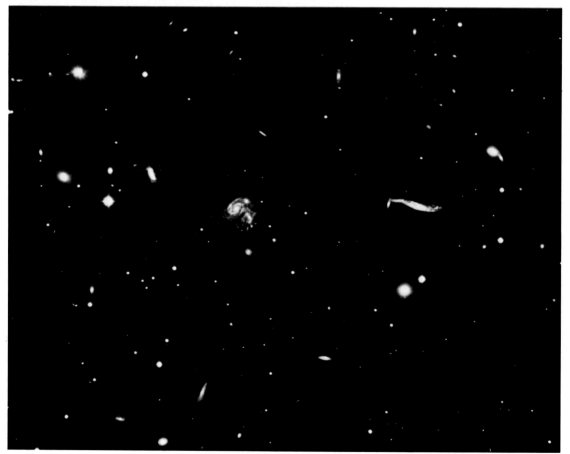

Milton Humason's skills took him from a job as mule driver to a staff position at Mount Wilson Observatory. Collaborating with Edwin Hubble, he spent 28 years recording the spectral patterns of distant galaxies. His work showing the lines had shifted toward the red end of the spectrum indicated that the galaxies were moving away from us. Hubble's concept of an expanding universe related red shifts to distance: The farther away a galaxy, the faster it was receding. Below, the vertical arrow indicates the position of an element, calcium, in the galaxy's fuzzy spectrum. Horizontal arrows show the calcium red shift in relation to lines of a standard spectrum.

CALIFORNIA INSTITUTE OF TECHNOLOGY (ABOVE); HALE OBSERVATORIES

RELATION BETWEEN RED SHIFT
AND DISTANCE FOR THREE GALAXIES

GALAXIES

RED SHIFTS

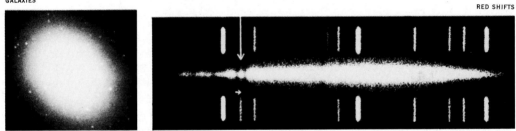

GALAXY IN VIRGO, 72,000,000 LIGHT-YEARS AWAY, RECEDES AT 720 MILES PER SECOND

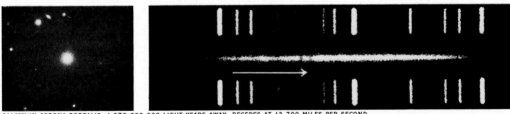

GALAXY IN CORONA BOREALIS, 1,370,000,000 LIGHT-YEARS AWAY, RECEDES AT 13,700 MILES PER SECOND

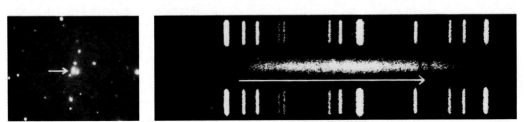

GALAXY IN HYDRA (ARROW), 3,800,000,000 LIGHT-YEARS AWAY, RECEDES AT 38,000 MILES PER SECOND

Thus, single-handedly, Shapley removed the sun and earth from a central position in the Milky Way and placed us close to its outskirts—an almost Copernican accomplishment. In the Missouri native's words, it was "a rather nice idea because it means that man is not such a big chicken."

Even before Shapley's work, astronomers were taking increasing interest in the cloudlike objects, or nebulae, seen throughout the sky. As spectroscopy, improving photographic techniques, and the great telescopes combined to give vastly better information, it appeared that these objects could be grouped in two broad categories. The Great Nebula in Orion, the Ring Nebula in Lyra, and the Crab Nebula were diffuse clouds of glowing gas. But many others, such as Andromeda, were spiral-shaped.

Shapley believed that all the visible universe, including both stars and nebulae, was in or on the edge of one great system, the Milky Way galaxy. His sharpest opposition came from Heber D. Curtis, classical scholar turned astronomer. The stage for Curtis's work had been set by George Ritchey of Mount Wilson Observatory in 1917, when he photographed the nebula identified as NGC 6946 (from Dreyer's *New General Catalogue)* and found a starlike object that had not been visible in earlier photographs. Ritchey recognized it as a nova—a brilliantly erupting star—inside the white nebula. He then searched the files of plates at Mount Wilson and discovered that two similar novae had appeared in Andromeda in 1909. Following Ritchey's lead, Curtis found three more examples in the plates at Lick Observatory. From the apparent brightness of these novae, and assuming they were comparable in absolute brightness to nearby novae, he conjectured they must be well outside the Milky Way.

When Maj. Edwin Hubble was mustered out of the U. S. Army in 1919, he promptly made his way to Mount Wilson and set to work studying nebulae. In 1923 he discovered some new Cepheid variables, and decided to try to determine the distance to Andromeda. By 1924 he was able to prove that it was far beyond the Milky Way (his estimate, though short of the actual 2.3 million light-years, was almost a million)—and, therefore, a galaxy comparable to our own. Hubble, at last, had discovered the universe.

STARS SHINE with characteristic colors that tell us how hot their surfaces are. When analyzed with a spectroscope, the light of a star not only shows distinctive lines characteristic of elements and compounds in its atmosphere; it also peaks in color intensity in the range that gives the star its dominant hue.

Color temperature is a familiar characteristic of hot objects (witness the terms red-hot and white-hot). If a piece of refractory material such as a ceramic is heated in a furnace, it glows dull red at about 3,000° C., cherry red at 4,000°, yellow at 5,000°, then bluish white. If the temperature continues to rise, the peak of the emission moves past the blue and violet and into the ultraviolet.

At the end of the last century, Edward Charles Pickering and his colleagues at Harvard College Observatory undertook to classify stars according to a simple sequence based on their spectral patterns: A for stars with strong hydrogen lines, down to Q when the lines were almost too faint to observe. In time some letters were dropped, new ones added, and the original alphabetical order altered. Today, students remember the Harvard classification sequence by the positive thought *Oh, Be A Fine Girl, Kiss Me.* Three subclasses were added to the cool red stars: *Right Now, Sweetheart.* O stars are very hot, some as high as 50,000°. The letters B, A, F, G, K, M represent progressively cooler stars down to about 3,000°.

If the spectroscopic lines of stars in nearby galaxies are compared with those of stars in the Milky Way, the agreement is very close. We can conclude that, on the average, the stars in both cases are made of the same elements in about the same proportions, although individual stars may exhibit marked differences.

As we turn our telescopes to more distant galaxies, however, a remarkable phenomenon becomes evident. Their lines retain the characteristic patterns of elements, but are systematically shifted toward the red end of the spectrum. The more remote the galaxy, the greater the "red shift."

This shift in light waves is comparable to the more familiar Doppler effect or change of frequency in sound waves from a moving source. When you listen to the wail of a train whistle or an emergency siren, movement of the approaching vehicle crowds up the sound waves so that your ears receive more per second; as the frequency increases, the pitch rises. Then, as the siren moves away, the waves stretch out and the pitch falls.

Similarly, light gains in frequency (is blue-shifted) when the source is approaching, and loses in frequency (is red-shifted) when the source is receding.

In 1912 at Lowell Observatory, Arizona, Vesto Melvin Slipher found that the Andromeda nebula—then still thought to be a cloud within our own galaxy—exhibited a blue shift of its spectrum: It was approaching us at a speed of 190 miles per second. That observation in itself was no cause for astonishment, for certain stars exhibited such velocities toward or away from us. But Slipher went on to measure other spiral nebulae, and found that the spectra of 11 out of 15 were *red*-shifted; furthermore, the largest red shifts corresponded to velocities of 685 miles per second, far greater than was typical of stars in the galaxy. As Slipher continued his work, the situation grew more extreme. Almost all his new measurements showed red shifts, and the fainter nebulae showed the greater velocities of recession.

When Hubble later proved that Andromeda and other spiral nebulae were, indeed, distant and separate galaxies, Slipher's observations became more acceptable.

If the nebulae were within the Milky Way, it was difficult to conceive why they should have such abnormal spectral shifts compared with stars; but if they were distant galaxies, the conflict disappeared.

Soon after Hubble embarked on his study of the distribution of galaxies, Milton Humason became his steadfast collaborator. They were a remarkable team. While Hubble gauged the distances of galaxies, Humason recorded their red-shifted spectra. To reach deeper into the universe, they had to target very faint galaxies. Obtaining their spectra required extraordinary skill and patience on Humason's part. A spectrum photograph is no simple snapshot. After selecting a faint nebulosity amidst thousands of sparkling stars, Humason had to center the image on the narrow slit of his spectrograph and hold it there from dusk to dawn. Every few minutes, the fine motions that tracked the telescope on the source needed a tiny adjustment to hold the spot of light in the middle of the slit. For the most distant galaxies, one night of photographic exposure was not long enough. After protecting the plate from stray light during the day, the exposure was resumed through the next night and sometimes for several nights.

Mountaintop temperatures brought a penetrating chill as the observer sat for hours at a time before the controls. With only infrequent breaks for coffee or a midnight lunch, he stayed at the telescope until the first hints of dawn. Night after night of such observing can build up a heavy burden of fatigue. Only the thrill of discovery that comes with scrutiny of the developed spectrum plate can compensate for the monotony of such long vigils.

By 1936, Hubble and Humason had undertaken the investigation of more than a hundred individual galaxies and ten clusters. The plot of red shift versus distance was a straight line; not a single galaxy proved an exception. Red shift was directly proportional to distance. Thus a galaxy at ten million light-years recedes at 100 miles per second; a galaxy at one billion light-years is racing away at 10,000 miles per second.

The cluster of galaxies called Ursa Major II contained the most distant galaxy they could then observe, and it was receding at 26,000 miles per second. (Just an hour from now, it will have rushed more than 90 million miles farther away from us.) The impact of such great velocities on some astronomers strained their belief in the meaning of the red shift. They speculated about "tired" light that had lost energy on its extremely long journey, and reasoned that the loss would be translated somehow into an increase in wavelength, a red shift; but no satisfactory theory has yet been offered for tired light.

Hubble's picture seems to place us at the center of an expanding universe, in which all other clusters of galaxies are moving away from us. But our central position is only an illusion. Consider a raisin yeast cake; let each raisin represent a cluster of galaxies, and imagine an observer on one raisin. As the cake bakes and swells, the raisins remain the same size, but they steadily separate. Every raisin moves away from the observer, and from every other raisin. The farther apart a pair of raisins, the faster they separate. When the cake has doubled in size, the distance between adjacent raisins has doubled.

No matter which raisin the observer is attached to, all others will move away. Merely because he sees all raisins moving away does not put him at the center, for his view from any other raisin would be the same. The cake has a center only because we can see its boundaries; but if we made an infinitely large cake, we could not find its center.

In an expanding universe, how do we account for *blue*-shifting galaxies—those that are approaching us rather than receding? The answer lies in the relative movements of galaxies within clusters. Andromeda, as part of our family of galaxies, can move toward the Milky Way even as the entire cluster is moving away from all

other clusters. Picture the railroad conductor who walks toward us from the front of the coach as the train carries us all away from the city.

As far as Hubble's telescope could probe, the proportional relationship between distance and velocity was constant. But the gravitational pull of every galaxy in the universe on every other galaxy—unless we accept the steady-state theory—must eventually slow down the expansion. How much greater was the rate of expansion in earlier epochs? We can determine that only if we can extend our investigations to the most remote galaxies and compare them with relatively close galaxies, to see how the relationship of red shift to distance changes with time. Distance measurement is at the heart of cosmology, but the problem is immensely difficult.

The giant pulsating Cepheids are hundreds to thousands of times as bright as the sun, yet Hubble could see them in only the nearer galaxies. With that base of distances, however, astronomers could calibrate other range indicators such as the brightest stars in galaxies, the infrequent novae and supernovae, great clusters of stars, and the large glowing gas clouds like the Orion nebula. With these new gauges, distances to more remote galaxies were estimated.

But toward the edge of the universe, all of these indicators become indistinguishable, and we must resort to measuring the light of the brightest galaxy within a cluster. Since galaxies themselves evolve in brightness, a way must be found to correct brightness for age. No precise measure has yet been found, but astronomers are searching diligently for new yardsticks in other portions of the electromagnetic spectrum.

Astronomer Jerome Kristian positions Palomar's 48-inch Schmidt Telescope. In the 1950's, Palomar—aided by a National Geographic Society grant—used the Schmidt to photomap the heavens for a 1,758-plate sky atlas.

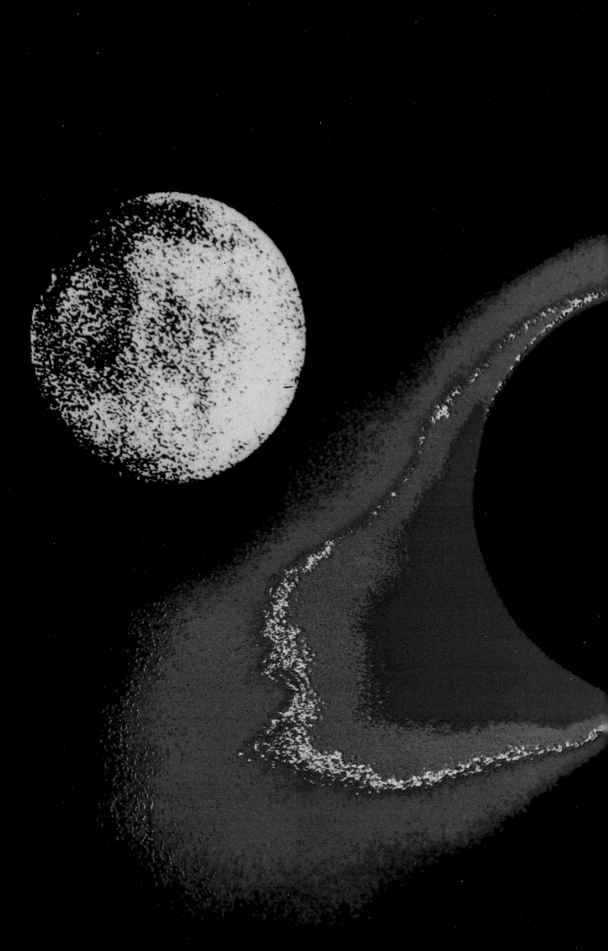

Our Star
the Sun

THE SUN IS OUR NUCLEAR power plant in the sky. It runs on a ten-billion-year supply of hydrogen fuel, producing a steady flow of heat and light that sustains all life on earth. At 8 cents per kilowatt hour, New Yorkers alone would have to pay 400 million dollars a day to provide the light that comes to them free from the sun.

Solar energy reaches our planet at the rate of five million horsepower per square mile. This steady flow is called the "solar constant"; no measurable change in it has been detected in more than a century of careful observations.

Under the protective blanket of the earth's atmosphere, we enjoy the benign heat and light of a seemingly peaceful sun; but take away the air—or even its trace of ozone—and the sun would be deadly. When astronomical instruments are carried above the atmospheric screen by rockets or satellites to measure solar emission of ultraviolet rays, X rays, and energetic particles, we discover that the sun is sputtering and erupting erratically. These ionizing rays tear electrons loose from the atmospheric molecules at heights above 50 miles and create earth's ionosphere, an electron mirror that reflects

Overleaf: Man-made eclipse, achieved by an instrument aboard Skylab, reveals the sun's corona, a vast halo of luminous gas. Scientists added colors to show variations of brightness. At this moment, the moon (left) was eclipsing the sun for earthbound observers.

HIGH ALTITUDE OBSERVATORY AND NASA
(SOLAR ECLIPSE: JUNE 30, 1973)

short-wave radio signals. Periodically the sun flashes a brilliant flare that releases energy comparable to that of billions of hydrogen bombs. Flares also generate solar cosmic rays—particles with energies up to billions of electron volts—that are a potential hazard for astronauts.

The impact of a solar flare on the ionosphere—and hence on radio conditions below—can be catastrophic. Short-wave transmissions black out immediately, compass needles swing wildly, circuit breakers trip on power lines, and teletype messages turn to gibberish. These disruptions may persist for days, yet we humans have no direct sensation of the commotion high above us. At most we may see great sheets of auroral lights rippling across the Arctic and Antarctic skies; and when the magnetic storm finally expires, the last residue of energy tints the heavens with a faint red air-glow that stretches from pole to pole.

A thin wind of hot solar gases, mostly hydrogen and helium, blows out of the sun continuously and sweeps past the earth toward the outer planets. When the sun is quiet, this solar wind may drop to a breeze of about a hundred miles per second, but at times of flare activity it gusts to more than a thousand miles per second. Large clouds of ionized gas called plasma then rip loose from the sun to drift off into space, spreading across a front of millions of miles by the time they reach the earth's orbit.

The vast flood of energy that pours into space from the sun originates in a thermonuclear furnace at its center. There, under incredibly intense conditions of temperature, pressure, and density, nuclear fusion converts hydrogen to helium, in the process releasing energy in the form of gamma rays and the elusive subatomic particles called neutrinos.

The temperature of the sun, about 5,700° C. at the surface, is 15 million degrees at its core. The density there is 12 times the density of lead, and the pressure is about a hundred billion times that of earth's atmosphere at sea level. It is difficult to believe that the core substance could be so concentrated and still behave like gas. The explanation is that collisions at the high central temperature are so violent that electrons are ripped off the atoms, leaving the nuclei bare. Normal atoms cannot press closer than about 1/100 millionth of an inch before their electrons touch; but when the electrons are stripped away, the remaining nuclei can approach a hundred thousand times closer. The interior of the sun is a violently agitated mob of electrons and naked nuclei streaking about at breakneck speeds.

Sir Arthur Eddington described the solar interior in vivid terms: "Crowded together within a cubic centimeter there are more than a [trillion trillion] atoms, about twice as many free electrons, and 20 [billion trillion] X rays. The X rays are travelling with the speed of light and the electrons at 10,000 miles a second. Most of the atoms are ... simply protons (*i.e.*, hydrogen nuclei) travelling at 300 miles a second. Here and there will be heavier atoms, such as iron, lumbering along at 40 miles a second. I have told you the speeds and the state of congestion of the road; and I will leave you to imagine the collisions."

Most of the mass of the sun is concentrated toward the center, about 90 percent within the inner half of its radius. Halfway out from the center, the density of the solar interior is about equal to water; at the surface, it has thinned out to about 1/10,000 of the density of our air.

All the warmth and light that flood through the sun's surface evolve from the moderated X rays that travel outward from its core. A photon, the unit of light or other radiant energy, takes only about eight minutes to travel from the surface of the sun to the earth. But that same photon originated in a gamma ray that degraded to an X ray, then careened, collided, and jitterbugged its way about the crowded solar interior for thousands or even millions of years before approaching the surface. Every collision

robbed the X ray of some of its energy.

A relatively cool outer layer sits over the sun's much hotter interior—a prescription for turbulence and convective heat circulation. Gas heated below expands and rises toward the top; at the surface it cools and turns back downward. Thus when we look at the surface of the sun with a telescope, it appears to be roiling with rising and descending columns of gas. The entire disk is mottled and granular, as though paved with cobblestones. Ranging from 150 to 1,000 miles across, the granules form and dissipate in about ten minutes. From a great distance the edge of the sun appears sharp, as though it marked a distinct surface; but in fact the solar surface is a layer of gas several hundred miles thick, called the photosphere.

The first clue to the sun's variability is the formation and disappearance of sunspots in the photosphere.

When the disk is dimmed by haze, especially near sunset, it is sometimes possible to distinguish large dark spots with the unaided eye. Long before the telescope was invented, the Chinese recorded sightings of such spots and referred to them as "flying birds." Galileo's description of the coming and going of spots, as seen with the aid of his telescope, contradicted Aristotle's ancient dogma that the sun was pure, unblemished fire, and many philosophers and churchmen of the time refused to use the instrument rather than risk being convinced.

Two hundred years ago the spots were thought to be solid mountaintops protruding above an ocean of glowing lava. In 1774, however, Alexander Wilson observed that the spots had inclined edges, like the slopes of a crater leading to a dark interior. Sir William Herschel went so far as to suggest that the sun's interior supported intelligent life.

Actually, sunspots appear dark only because of contrast; they are 2,000° cooler than the bright photosphere. If all the photosphere were *(Continued on page 58)*

New Light on the Sun

Man has worshiped or studied the sun ever since he first recognized its importance to his existence. Yet only recently have scientists begun to penetrate the mysteries of the fiery, gaseous sphere whose light and heat account, directly or indirectly, for almost all the energy used by man. Increasingly refined knowledge of solar composition and nuclear reaction rates, combined with data from sophisticated new instruments, now enables astrophysicists to theorize in detail on both the interior of the sun and its atmosphere—as depicted in this artist's conception (with colors arbitrarily assigned for contrast).

At a temperature of 15 million degrees, the central *core* remains gaseous despite a density 12 times that of lead. By nuclear fusion this solar furnace converts hydrogen to helium, releasing energy in the form of gamma rays. These quickly degrade to X rays, whose high-energy photons make their way toward the surface through the *radiation zone*, losing energy as they repeatedly collide with ions and electrons and transforming into light and heat. In the *convection zone* currents of gas evolve, flow upward, release energy at the much cooler surface (about 5,700° C.), then return downward to be heated again. This vertical circulation of the currents gives the surface a rippling, mottled appearance; the resulting granules constantly form, dissipate, and reform.

The surface, or *photosphere,* actually comprises a layer of gas several hundred miles thick, pockmarked by large, slowly drifting *sunspots.* The *active regions*—bright areas of intense magnetic activity and highly concentrated convection currents—give rise to graceful *magnetic arches* of spiraling gas. Disturbances in the photosphere sometimes trigger the eruption of *solar flares* along these arches. Out of the photosphere and far into space blows a continuous *solar wind,* which sometimes gusts to a thousand miles a second.

Above the photosphere rises the thin shell called the *chromosphere,* briefly glimpsed as a red ring at times of total solar eclipse. Myriad fountainlike jets of gas, called *spicules,* shoot thousands of miles high through the chromosphere. Great *prominences,* clouds of incandescent gas, hang above the chromosphere and sometimes erupt violently when jolted by rapidly expanding magnetic fields. The thin, pearly *corona* surrounds the sun and sends out far-reaching streamers. Near the poles, prevailing *coronal holes* show that magnetic activity there is slight. Strangely, the temperature of the solar atmosphere increases rather than decreases with altitude, reaching more than a million degrees in less than 10,000 miles—presumably the result of frictional dissipation of shock waves from below.

SOLAR WIND

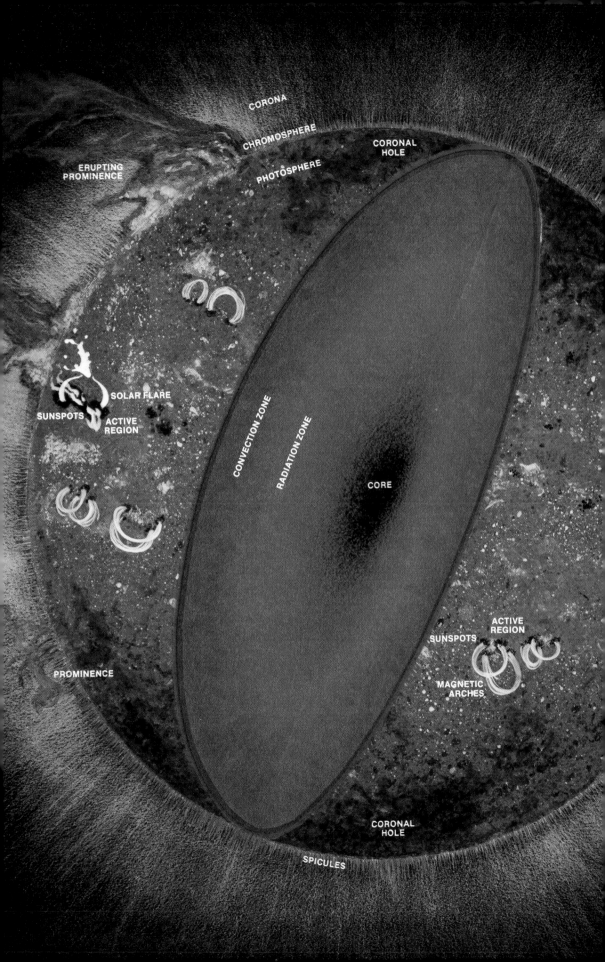

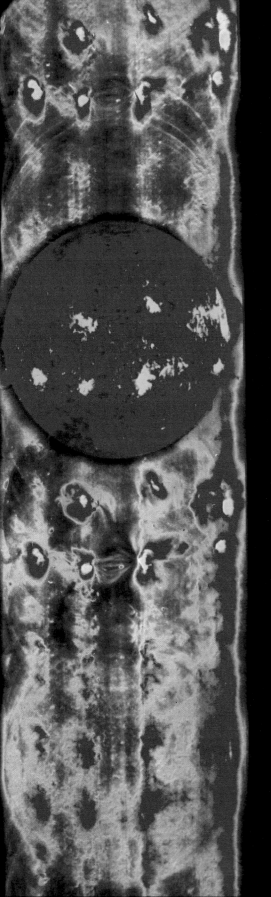

Investigation of the sun has progressed rapidly during the last century with development of various instruments for recording and analyzing its electromagnetic radiation. The photographs on these two pages, taken on September 5, 1973, during Skylab's second mission, show how the same active regions appear when recorded by different techniques.

Utilizing a grating between sun and film, the spectroheliograph can form various images of the sun by selectively registering the rays of individual elements or ions. With increasing temperature, elements lose more and more electrons; each stage of this ionization, photographed at its own wavelength, reveals details of composition and temperature at a different level of the solar atmosphere.

To make the series of images at left, the spectroheliograph grating dispersed light into many different wavelengths. The image of $100,000°$ helium in the sun's chromosphere appears as a mottled disk, and shows the chromospheric network and bright active regions. Under it, and reproduced in the photograph below, the image in the light of the ionized element iron-XV indicates million-degree zones in the corona, with magnetic arches bridging and interconnecting the active regions.

NAVAL RESEARCH LABORATORY AND NASA

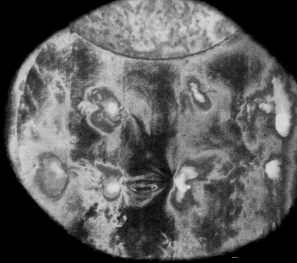

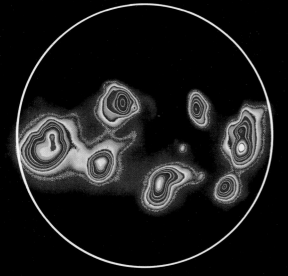

Solar X-ray telescope records magnetic arches in the corona. Magnetic fields of the active regions extend from the photosphere into the corona, trapping superhot gases. These take the form of feathery arches whose bases are rooted at the photosphere in areas of opposite magnetic polarity. At the poles, cool "holes" in the coronal gases appear as darker areas where magnetic activity is slight.

AMERICAN SCIENCE AND ENGINEERING AND NASA

Photograph in shorter-wavelength X rays than the one above shows only the hottest areas of the active regions. Astronomers arbitrarily assigned the colors to distinguish contours clearly. The magnetic fields dictate the size and shape of X-ray-recorded features concentrated toward the equator.

MARSHALL SPACE FLIGHT CENTER, THE AEROSPACE CORPORATION, AND NASA

Scanning the sun's photosphere through a polarizing filter in a study coordinated with Skylab's efforts, a ground-based magnetograph sends data on magnetic fields in the active regions to a computer. Translated to a television pattern and photographed, the image appears at left. Red indicates areas of magnetic force directed upward, and blue, downward—base points of the magnetic arches.

KITT PEAK NATIONAL OBSERVATORY

JAY M. PASACHOFF

Inscription "Take me to your leader" marks one of six Nike-Asp rockets awaiting launch during the total eclipse of October 12, 1958. The missiles blasted aloft from the U.S.S. Point Defiance in the South Pacific in the first application of rocket astronomy to the study of a solar eclipse. United States Naval Research Laboratory scientists, under the direction of author Herbert Friedman, used the rockets to reach above the obscuring effects of the atmosphere. Their instruments clearly detected X-ray activity in sunspots and in the corona. A pioneer in rocket astronomy, Dr. Friedman (below) explains a map indicating diffuse X-ray sources in the universe. Above, during a more casual observation of the sun, children of a mission school in Kenya look through special filters set up on the shore of Lake Rudolf by visiting scientists during the total eclipse that occurred on June 30, 1973.

NATIONAL GEOGRAPHIC PHOTOGRAPHER JAMES P. BLAIR (BELOW); UNITED STATES NAVY (OPPOSITE)

obliterated by one huge sunspot, the light cast would still be comparable to the typical sunlight of late afternoon.

The number of spots on the sun waxes and wanes in cycles that average about 11 years, and near a cycle's minimum the disk may be completely devoid of them. Spots move from east to west with the sun's rotation, but their motion tells us that the sun rotates in a very peculiar way: The gaseous sphere twists about its axis, so that the rotation period is only 25 days at the equator but 29 days at the poles. Sunspots also show a slow drift in latitude.

The spots usually occur in what are termed the sun's "active regions"—areas of intense convective and magnetic activity that create bright patches called plages. In fact, the most significant physical property of sunspots is the intensity of their magnetism, many thousands of times greater than that of the earth. Every aspect of solar activity seems to be tied to variations of the magnetic fields.

These powerful fields may originate deep in the interior. Because of the solar rotation, the ascending and descending streams of gas get twisted into whirls which detach like smoke rings and rise toward the surface, where they finally break through to form pairs of spots.

IN 1889, WHEN HE was only 21, George Ellery Hale developed the spectroheliograph, an instrument with the ability to narrow down the range of color to a single spectrum line and to produce an image in that wavelength. It was possible, for example, to provide an image of the sun in the red light of glowing hydrogen or the violet light of calcium, or in other colors of the various elements that are excited in the solar atmosphere. By isolating each element, fine details are revealed that are otherwise blurred in the white-light tumult of all the combined elements.

By 1936 Robert R. McMath and his colleagues at the University of Michigan had brought Hale's instrument to a high degree of perfection, and produced a remarkable movie of the active sun. At close range it exhibits a bland face, pockmarked by the sunspots. With increasing altitude, the picture changes to marvelous whirlwinds and snaking patterns. Flaming streamers sometimes lick upward for tens of thousands of miles, or lie over horizontally like a vast prairie fire fanned by powerful winds. Great prominences hang a hundred thousand miles high in luminous arches, raining downward toward the surface.

A few thousand miles above the photosphere, the solar atmosphere is so thinned and transparent that it becomes virtually invisible in the glare of light from the photosphere. But when the brilliant disk is masked by an eclipse, we discern a very interesting profile. Then the sun's corona, or extended halo, becomes briefly visible in one of the most dramatic of all nature's spectacles.

The earliest historical record of an eclipse goes back more than 4,000 years to October 22, 2137 B.C., and appears in the Chinese classic *Shu Ching*. Predicting eclipses was obviously a hazardous occupation, for the book contains this decree of the emperor:

"Being before the time the astronomers are to be killed without respite; and being behind the time they are to be slain without reprieve."

The thrill of observing an eclipse has long tempted astronomers to the far ends of the earth. In the days of sailing ships, one Scottish scholar had the unenviable record of traveling 75,000 miles to vantage points for six eclipses and actually seeing only one; clouds interfered on all other five occasions. The French astronomer Pierre Janssen was so intent on photographing the eclipse of 1871, during the Franco-Prussian War, that he risked German rifle fire to escape the siege of Paris in a balloon. But when he reached the African coast and the eclipse path, rain hid the event.

Fortunately, not all eclipse expeditions end up in such frustration. In 1842, astronomers in southern France were the first to take careful note of the very faint and extended outer atmosphere of the sun. As the moon blocked out the brilliant disk, a pearly white corona with delicate streamers and exquisitely curved arches stood clearly revealed. Close to the black edge of the moon a reddish ring, the chromosphere, encircled the sun; above this layer, the luminous red clouds of gas called prominences reached high into the corona.

As we have seen, the sun's temperature decreases steeply from the thermonuclear furnace's 15 million degrees to about 5,700° at the surface. In the outer atmosphere we would expect even cooler gas, but instead the temperature shoots above a million degrees in less than 10,000 miles—the thickness of the reddish layer called the chromosphere. Normally, heat flows from a hot area to a cool one; how, then, can the chromosphere and corona derive their high temperatures from a much cooler photosphere? Obviously it is not by conduction or radiation, as the kitchen burner heats a kettle. Some mysterious form of heat transfer feeds energy directly into the high atmosphere from far down in the photosphere. Martin Schwarzschild of Princeton University has suggested that the bubbling granules break like ocean waves against the bottom of the chromosphere and create a tremendous, rumbling roar. As these sound waves rush upward into the more rarefied gas, he reasons, they increase in speed until supersonic shocks occur, like the crack of a whip; then the shock-wave energy dissipates in friction, heating the gas of the corona higher than a million degrees.

Pictures of the rim of the sun just above the granulation show a fountainlike structure: Tongues of gas called spicules shoot like jets above the bursting granules. They surge up and fall back again in five to ten minutes, rising with speeds of 10 to 15 miles a second to heights as great as 6,000 miles.

The chromosphere has been called the "spray of the photosphere," since it appears to be composed almost entirely of spicules.

Where the chromosphere blends into the corona, the solar gas suddenly becomes about a thousand times thinner. So transparent is the corona that stars can easily be seen through it at the time of an eclipse, and comets brush by with little noticeable effect. If the eclipse occurs at the sunspot cycle's maximum, the corona is shaped symmetrically, like a large dahlia with a black moon at the center. At sunspot minimum, great equatorial streamers distort the symmetry and can be seen stretching millions of miles into space. Indeed, the outer fringes of the corona overlap the earth and possibly reach to the most distant planets.

In 1930, Bernard Lyot of France solved the problem of observing the corona without the aid of an eclipse. He constructed the coronagraph, a telescope with an internal occulter—a disk that substitutes for the moon in blocking out the orb of the sun. With the coronagraph we can see huge clouds of bright gas reaching as high as a hundred thousand miles. Where the gas clouds are anchored to the photosphere, violent convection twists and shifts the ropes of plasma about, causing the material in the high arches to react in spectacular whipping, streaming, and eruptive patterns.

In visible light, the thin gas of the chromosphere and corona shines feebly compared with the bright photosphere, but in far ultraviolet and X rays, the pattern is reversed: Then the photosphere appears as dark as a sunspot, while the emission from the chromosphere and the corona shines out brightly.

This is where rocket and satellite astronomy assumes an important role.

When captured German V-2 rockets were brought to the White Sands Missile Range in New Mexico after World War II, their warheads were replaced by scientific payloads, and among the first instruments to be carried were (Continued on page 64)

JOHN LUTNES, KITT PEAK NATIONAL OBSERVATORY

Bringing the Sun Down to Earth

Navajo Indian students on a tour of Kitt Peak examine a 30-inch image of the sun within the McMath Solar Telescope. Dark glasses shield their eyes from the reflection on the viewing screen. The McMath, world's largest solar telescope, stands atop a high ridge of the mountain (opposite). The vertical tower rises 110 feet and supports the heliostat, or tracking mirror, which reflects the sun's light down a diagonal, 500-foot-long tunnel. An image-forming mirror at the end of the tunnel sends the light back up to a ground-level mirror; this third mirror projects the beam to laboratories and to the observing room. The combination of mirrors can direct light to various stationary instruments. Astrophysicist Leo Goldberg, director of Kitt Peak since 1971, sits at a 24-inch ruling engine used in making diffraction gratings for spectral study. A grating—a series of close, parallel grooves ruled onto a thin metallic layer on a glass plate—disperses solar and stellar light into various colors, and reveals greater spectral detail than a prism.

J.R. EYERMAN (ABOVE); N.G.S. PHOTOGRAPHER JAMES P. BLAIR

SOLAR IMAGES BY NAVAL RESEARCH LABORATORY AND NASA, CORONAL IMAGES BY HIGH ALTITUDE
OBSERVATORY AND NASA, COMPOSITES BY NAVAL RESEARCH LABORATORY; SKYLAB 3, NASA

Arcing into the sun's atmosphere, an erupting prominence glows with colors artificially added to the photograph: White indicates areas of most intense radiation, red the areas of least intensity. Eruptions such as this create disturbed clouds of gas in the corona, known as loop transients. Seen only by masking the bright sun, the loop transients appear as filmy white arches in two composites (facing page) timed at a 30-minute interval. To show the origin of the loop transient, the sun and spurting prominence appear superimposed on the black mask. With the sun's brilliant light eclipsed, the star Regulus reveals its presence. Scientist-astronaut Edward G. Gibson (opposite) works at the complex control panel for Skylab's solar telescopes.

ultraviolet spectrographs. In 1946, the first solar spectrum above earth's atmosphere was photographed by a team of scientists from the Naval Research Laboratory, led by Dr. Richard Tousey. Over the years that followed, the technology of rocket astronomy made steady progress. Photometers — instruments for measuring light intensity — and spectrometers eventually reached to the X-ray limit of the sun's radiation, so that by the time of the first observatory satellite the general features of the quiet sun were well mapped. Solar rocket astronomy played a large role in the entire development of astronomy from space platforms, and I often recall some of the more exciting occasions in my own experience during that period.

BEFORE THE DISCOVERY of solar X rays, theorists struggled to explain radio fadeout as a consequence of a flash of intense ultraviolet radiation in a solar flare. From our results with rocket measurements of X rays and ultraviolet rays, it seemed clear to my colleague Dr. Talbot Chubb and me that the cause of radio fadeout must be an X-ray flash. To test our hypothesis, we had to find a way to launch a rocket within the first few minutes of a flare outburst. The liquid-fueled Aerobee rockets that had been flying from White Sands had to be pressurized with helium before firing; they could not be held in the launching tower ready to go at some uncertain future moment when a flare might erupt. The problem was solved by hanging an inexpensive solid-propellant rocket, the Deacon, from a Skyhook balloon and releasing it at sea. When a flare began, we could send a firing command to the rocket, which would promptly spurt upward into the ionosphere. By launching the balloon early in the morning, we could have nearly all day to await a flare. If none occurred, the rocket was fired as a precaution against its drifting back over land.

From San Diego in August 1956, we went 500 miles out to sea with two vessels, the landing ship dock U.S.S. *Colonial* and the destroyer U.S.S. *Perkins.* The *Colonial* was a floating machine shop and laboratory, with a helicopter deck from which we could launch balloons.

By cruising at just the right speed to match the wind, it was possible to inflate a balloon in a straight-up position. At launch, it stood a hundred feet above the deck with a 20-foot-wide bulge at the top, a giant onion of polyethelene plastic. At an altitude of 80,000 feet, the gas would expand the plastic bag to 30 times its initial volume.

Dangling from the balloon was a 100-foot nylon rope to which we attached radar reflectors and, below them, the instrumented rocket rigged in firing attitude. In 90 minutes the combination rocket and balloon, dubbed "Rockoon" by one of its inventors, Dr. James Van Allen, had soared to 80,000 feet and was racing westward at 30 knots. The *Colonial* lumbered after it at 15 knots, while the *Perkins*, with a speed of 26 knots, stayed close on the tail of the balloon while tracking it with radar. The telemetry receivers aboard the *Colonial* could read the rocket's wireless signal up to 125 miles.

We started the cruise with ten Rockoons. Each morning we launched early, then waited while the Rockoon and the ships played fox and hounds. On the fourth day, just at lunchtime when only Robert Kreplin and William Nichols were tending our command station, Richard T. Hansen, on duty at the University of Colorado's High Altitude Observatory at Climax, saw a flare begin to form.

As Kreplin overheard him describe the flare on the radio, he quickly decided to push the firing button. The rocket reached altitude while the flare was near its peak, and recorded an intense flash of X rays — exactly what theory needed to explain the radio fadeout phenomenon.

The proof of X-ray emission from a flare spurred a great program of further studies during the International Geophysical Year of 1957-58. After the Rockoon expedition, a two-stage vehicle was developed by mating the Deacon with a Nike booster which replaced the balloon. Our scene of operations shifted to the island of San Nicolas off the coast of southern California. There we mounted our simple rail launcher on a rise looking out over a quiet beach, where hundreds of seals played in the sun while we patiently waited for solar flares. A number of large flares were observed in 1957 and 1958, and the full power of the solar outbursts became increasingly clear.

The year 1957 brought Sputnik and ushered in the satellite era of space research. One satellite observing the sun continuously was worth hundreds of individual rocket shots of a few minutes each. Efforts to launch Vanguard satellites equipped with X-ray detectors were unsuccessful; but finally, in 1960, the Navy's Solrad program got off its first successful solar X-ray monitor.

The initial Solrad mission was followed in 1962 by the first of NASA's series of Orbiting Solar Observatories. These OSO spacecraft carried increasingly complex spectrometers for the entire ultraviolet and X-ray range of the spectrum. Since then the program has steadily increased in sophistication, and the Solrad 11 mission is scheduled late in 1975 to place a pair of satellites 180° apart in 70,000-nautical-mile orbits to monitor the sun continuously with a large complement of photometers.

But before moving into the new era of satellite astronomy, my colleagues and I had had a taste of adventure in the classical tradition of studying the solar corona at eclipse —except that we used rockets.

That the solar disk must radiate unevenly throughout the entire ultraviolet and X-ray spectrum was inferred by radio physicists long before rockets and satellites could place instruments above the obscuring air. Our success with the two-stage solid-propellant rockets in the Geophysical Year solar-flare studies encouraged an attempt to study the distribution of X-ray and ultraviolet sources over the face of the sun during the total eclipse of October 12, 1958.

Solar astronomer at Kitt Peak, A. Keith Pierce works with a spectrograph in the observing room of the McMath Telescope. Separating solar radiation into wavelengths characteristic of specific chemical elements, the spectrograph makes it possible for astronomers to study the sun's composition.

Twisting strands of gas break away from the sun's rim as a prominence erupts through the solar atmosphere. The powerful blast of energy on December 19, 1973, marked one of the mightiest eruptions in 25 years. Regions of strong magnetic fields appear bright yellow in this color-enhanced ultraviolet photograph made on Skylab's third mission, which ended in February 1974. When pressure in the sun's active regions builds sufficiently, eruption takes place. When it occurs near a prominence already hanging high above the surface, such a blast often propels a gas cloud outward at a speed great enough to escape the sun's gravity. Fewer eruptions develop toward the poles, where the predominance of darker areas indicates little magnetic activity.

NAVAL RESEARCH LABORATORY AND NASA

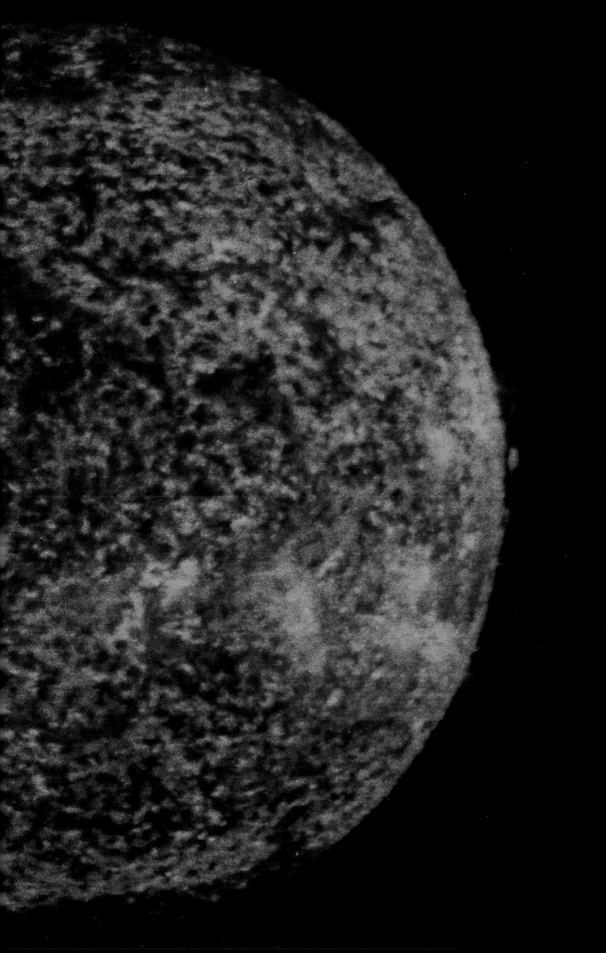

Hunting for neutrinos, energetic particles from the sun, Raymond Davis, Jr., sets his trap — 100,000 gallons of cleaning fluid — deep in a South Dakota gold mine. By capturing the elusive particles, believed to come from fusion in the sun's core, he hopes to prove how the sun yields energy. Dr. Davis uses the mine to shield his experiment from unwanted cosmic rays. Neutrinos can pass through the earth unhindered, but some should combine with an isotope of chlorine in the cleaning fluid to form a radioactive gas. At right, Davis adds liquid nitrogen to a charcoal trap designed to separate out the radioactive gas. So far the experiment has failed to identify any solar neutrinos. If further refinements don't find the missing particles, scientists may have to rethink their theory of nuclear fusion in the sun.

The plan was simple. Rockets would be fired in sequence as the eclipse progressed. If X rays came from a few localized active regions, the rocket instruments would register abrupt decreases in X-ray intensity as each center was blocked out by the moon.

The long track of the eclipse across the Pacific Ocean missed all of the South Pacific islands except for a few tiny coral atolls. Our expedition was organized as a joint venture of ground-based astronomers under the leadership of Dr. John Evans of the Sacramento Peak Observatory in New Mexico and a team of rocket astronomers from the Naval Research Laboratory under my direction. Dr. Evans' team chose Puka Puka in the Danger Islands as their base. To support the expedition, the Navy provided the landing ship dock U.S.S. *Point Defiance.*

Exotic South Sea islands, Polynesian natives, a naval warship, and modern rockets were a sufficiently unusual mix of ingredients to create a highly memorable experience. When our huge vessel dropped anchor off Puka Puka, we were greeted by a throng of natives who had never seen a ship larger than a copra schooner, or anything more mechanically complex than a bicycle. Yet within a few days the natives accepted us, our helicopter, and our other contraptions as though everything were quite normal.

While the astronomers set up their base ashore, we mounted six Nike-Asp rockets on the helicopter deck on the ship. The slim, 27-foot-long rockets pointed skyward like sleek arrows, poised to trace a trajectory with a peak of 150 miles.

On the day of the eclipse, clouds gathered over Puka Puka while the *Point Defiance* cruised under clear skies only 20 miles away. By eclipse time the rain came pouring down on the island, and the astronomers were completely washed out. But aboard ship the rockets departed on schedule, each with a shattering roar and a brilliant burst of flame. Soon after the last launch, the eclipse shadow reached the ship, and we enjoyed the beautiful spectacle of the corona.

A quick scan of the telemetry records showed that our rocket observations had succeeded. X rays came from high in the corona; even with the disk totally obscured, 13 percent of the X-ray emission remained unaffected. As the moon intercepted individual sunspots, the X-ray intensity diminished abruptly, showing that the sources were concentrated directly above the sunspot groups.

Pleased as we were, our joy was tempered by compassion for our astronomer friends on the island. A year's preparation before embarking and seven weeks of effort on Puka Puka had come to naught.

From 1958 on, great emphasis was placed on developing instruments to capture images from solar X rays and ultraviolet rays. In 1960, a simple pinhole camera riding an Aerobee rocket showed that as much as 80 percent of the sun's X-ray emission came from highly condensed parts of the corona, which covered only 20 percent of the projected area of the disk. From that modest beginning solar photography has made enormous strides, leading finally to the high resolution X-ray and ultraviolet cameras that flew on Skylab in the Apollo Telescope Mount.

The ATM offered all the capacity that contemporary solar-physics instrumentation for space was prepared to exploit—along with two scientist-astronauts. This teaming of scientists in space with scientists on the ground proved a remarkable success.

Each evening, experts at the Johnson Space Center at Houston evaluated performance as relayed from Skylab and formulated an operating plan for the next day. The scientist-astronauts aboard Skylab followed these plans, but were permitted a large measure of individual selection and judgment. Some discoveries were made purely on the basis of their observations. The total bag of short-lived phenomena, such as flares and eruptive prominences, satisfied the fondest hopes of the experimenters—160,000 solar images!

The enormous return of scientific data will require years of study, but a number of interesting results were quickly apparent. ATM's combination of spectroheliograph and coronagraph has linked the full chain from the surface far into the corona with exquisite detail. The images show that the hot coronal plasma is tied to the sun by closely knit magnetic loops that cover much of the disk. Where these loops are missing, large holes appear in the corona. Here, we suspect, are the routes by which the solar wind escapes the magnetic traps.

The corona is in constant motion. Great transient prominences, huge expanding loops, bubbles of billowing gas, and far-reaching spikelike streamers portray plasma escaping into space or falling back to the sun. Sharp surges break out of the disk to penetrate coronal forms and set off violent transformations a million miles on high.

Even the familiar spicules, those jetlike sprays of the chromosphere, appear in a new light in ATM photographs. Images produced in the wavelength of ionized helium show spicules of a giant variety, taller and broader and lasting as long as 30 minutes. The ATM images show them clearly over prevailing coronal holes near the poles, reaching to heights of 25,000 miles.

Small X-ray hot spots speckle the sun's surface; they last about eight hours and often flash microflares. They may account for much of the variation in the solar wind.

WE HAVE TALKED of mountaintop observatories and telescopes in space, but some of the most interesting solar astronomy of recent years has been conducted far below the surface of the earth.

The particle called a neutrino, produced in the process of nuclear fusion, has (like a photon) no measurable mass or electric charge, but it travels at the speed of light. It is perhaps the most elusive of all the known members of nature's subatomic zoo, passing through matter so easily that it escapes the sun unhindered and can roam the universe almost endlessly without being caught. If we could make an efficient neutrino telescope, we could investigate the very center of the sun, and the hottest stars in the galaxy would stand out more conspicuously than they appear through optical telescopes.

Deep in a gold mine under the Black Hills of South Dakota, neutrino astronomy has become somewhat more than a dream. There Raymond Davis, Jr., of Brookhaven National Laboratory uses a tank filled with a hundred thousand gallons of tetrachlorethylene cleaning fluid as a giant neutrino trap. It is far underground because the detector must not be confused by subnuclear particles produced by cosmic rays.

When a neutrino combines with an atom of the isotope chlorine-37, they form an atom of argon-37, a gas that can be removed by bubbling helium through the chlorine compound. The argon is then separated from the helium by trapping it in very cold charcoal. The radioactive argon can be identified by sensitive instruments.

Davis hopes to duplicate this laboratory reaction with solar neutrinos in his huge tank of cleaning fluid. With two million trillion trillion atoms of chlorine-37 on hand, his most optimistic expectation is about one neutrino a day. So far, he has not claimed to capture any solar neutrinos; the amount of argon-37 observed could be attributed to other factors. But the experiment is being steadily refined.

If the cleaning-fluid trap produces no neutrinos, the mystery of the missing particles may force solar physicists to reexamine their theories of the sun's internal activity. Meanwhile, Ray Davis continues his work with a monumental patience.

Cloud of incandescent gas, a coronal loop thrusts 130,000 miles above the sun's edge. Skylab astronauts took this X-ray photograph, computer-contoured and colored to distinguish radiation intensities. White shows highest intensities, red the lowest.

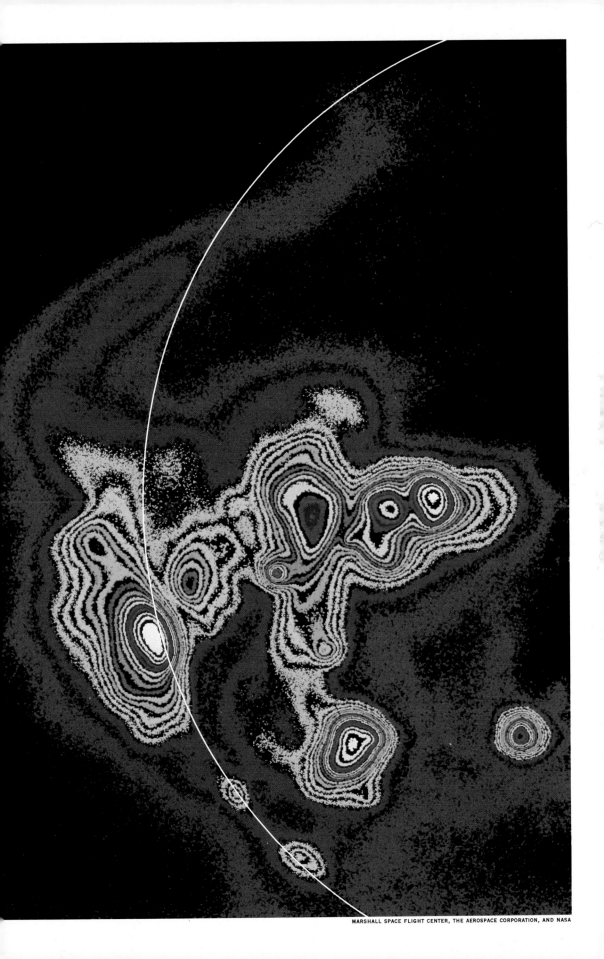

MARSHALL SPACE FLIGHT CENTER, THE AEROSPACE CORPORATION, AND NASA

The Family of Stars

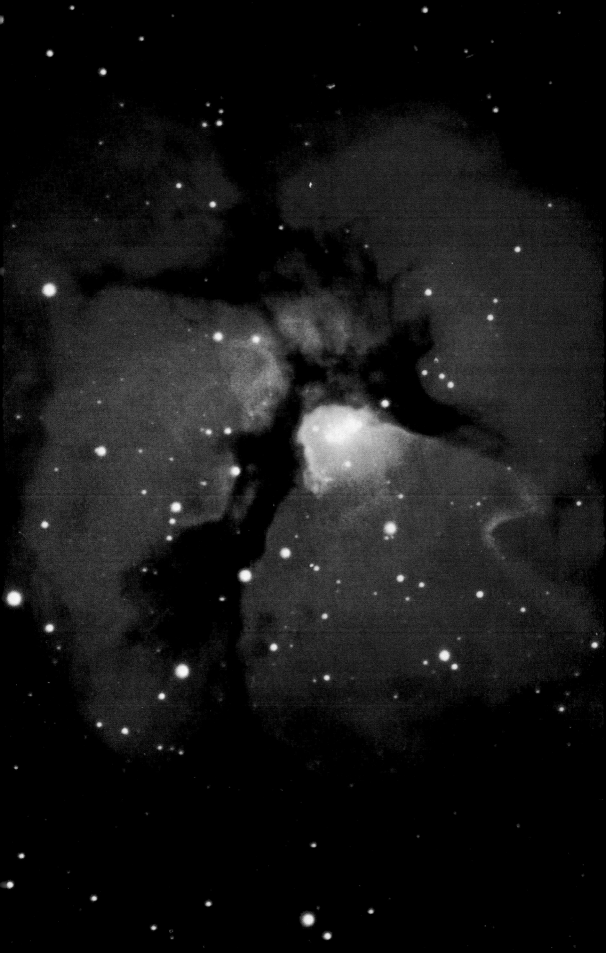

Most MEMBERS of the family of stars bear a close resemblance to the sun, but the distant relatives can be incredibly brighter or dimmer, bigger or smaller, heavier or lighter. Not only can one star vary greatly from another in luminosity, temperature, size, and mass; but a given star itself changes drastically as it evolves through its life cycle—and astronomers continue to discover strange and unforeseen stages in that cycle. Thus we have a still-growing vocabulary of evocative stellar terms, including red giants, blue supergiants, and white dwarfs; novae and supernovae; X-ray stars, neutron stars, pulsars, and black holes.

The life of a star begins by accident. In interstellar space, gas and dust are typically found at densities of a few particles per cubic inch. By chance collisions of the particles and by gravitational attraction, material accumulates over millions of years into a large, cold cloud. As it grows, condensations occur that become embryonic "protostars." In the Milky Way's luminous nebulae—enormous clouds of dust and gas, such as the Great Nebula in Orion and the Rosette Nebula—radiation pressure and shock waves from nearby stars are believed to speed up the formation of protostars. These nebulae are busy breeding grounds, studded with bright new stars, and diffuse protostars radiating in the invisible infrared range.

As gravity continues to act on a protostar, it grows increasingly dense. Matter falls faster and faster toward the center and heats up. A protostar containing about the same amount of material as our sun shrinks from a diameter of trillions of miles to the present size of the sun in about ten million years. By that time the internal temperature has risen to perhaps ten million degrees, and protons—hydrogen nuclei—collide with enough energy to overcome the repellent electrical force that normally keeps them apart. The force of such collisions causes hydrogen atoms to combine to form helium; with this process of nuclear fusion, the stellar furnace begins to burn.

Some 20 million years later, the star stabilizes: The outward pressure created by the heat of nuclear reaction is exactly balanced by the inward pull of gravity. For the next ten billion years the star will remain in this steady state. Our own middle-aged sun is now five billion years into this cycle and has about five billion to go.

Cosmologist George Gamow once remarked that the study of human population is much simpler than that of stars since, "whereas all humans have approximately the same life expectancy, the life expectancy of stars varies as much as from that of a butterfly to that of an elephant." The more massive the star, the shorter is its life span. Stars having much more mass than the sun burn their fuel prodigally in a flaming youth and plunge to an early death. For example, a star ten times as massive as the sun radiates a thousand times as much power, but survives for only a hundred million years. To sustain its greater weight, its nuclear furnace must burn hotter. The rate of nuclear fusion is very sensitive to temperature; at 20 million degrees the nuclear fuel is consumed 30,000 times faster than at 10 million degrees.

SCALE: SEE PAGE 4

Overleaf: Streaks of interstellar dust vein the reddish glow of the Trifid Nebula. Starlight transforms a dark cloud to luminous blue. Such celestial regions—birthplaces of stars—provide clues to stellar evolution. KITT PEAK NATIONAL OBSERVATORY

Smaller stars, on the other hand, are the Methuselahs of the stellar community. A star of one-tenth the mass of the sun can burn for a trillion years—or more than 60 times as long as the estimated present age of the universe!

Our sun began to shine billions of years after the first stars had formed in the Milky Way. By then, many of those early massive stars had already completed their relatively short life cycles and exploded their ashes into space. From such debris, combining with interstellar matter in a continuous pattern of cosmic recycling, new stars and planets are eventually born with all the elements necessary for the evolution of life—such as the iron in our blood and the calcium in our bones.

In a day of atomic weapons and nuclear power, it is sometimes difficult to remember that only 50 years ago the source of the energy of the stars was still a mystery. In 467 B.C., the philosopher Anaxagoras surmised from study of a large meteorite that fell in Greece that the sun was a mass of molten iron. Twenty-three centuries later, Lord Kelvin—whose name has been given to the absolute temperature scale—suggested that energy from the impact of meteorites falling into the sun was converted into heat and light. Kelvin and physicist Hermann von Helmholtz subsequently argued that the sun got hot by slowly shrinking—that is, by the transformation of gravitational energy into heat. They calculated that the sun therefore had been shining at its present rate for a hundred million years. But when the earth was found to be at least several hundred million years old, it became clear that the sun's gravitational energy was grossly inadequate to account for the light it had emitted over its lifetime.

The unknown source of stellar energy presented a serious problem for astrophysicists studying the internal structure and evolution of stars. In 1920 Arthur Eddington proposed that some kind of atomic energy *(Continued on page 84)*

THOMAS NEBBIA

Probing the interior of stars mathematically, Sir Arthur Eddington (below) computed the structure and temperature of stars and suggested correctly that atomic energy might cause stars to shine. Studies of white dwarfs led Jesse L. Greenstein (above) to conclude that these dying stars have exhausted all possibility of generating nuclear energy.

ROYAL ASTRONOMICAL SOCIETY

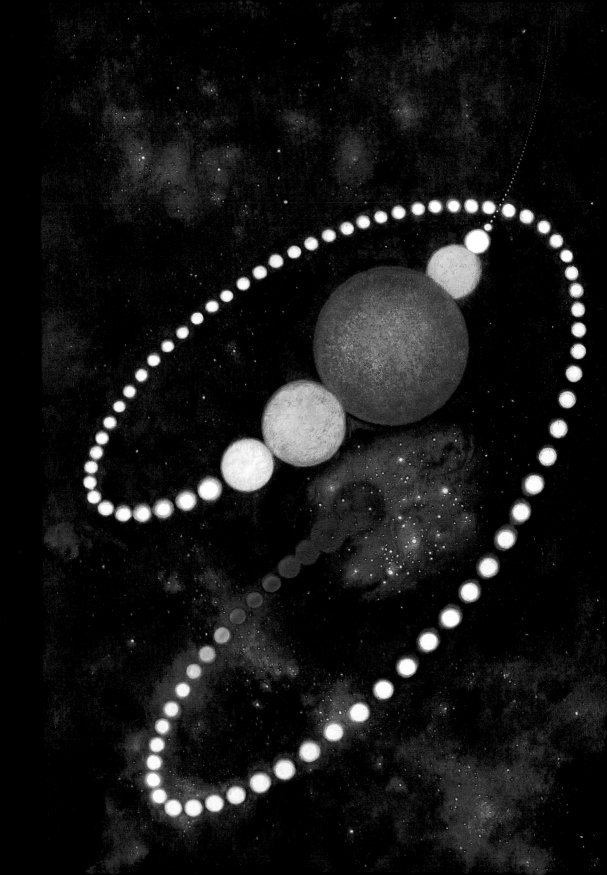

PAINTING BY DAVIS MELTZER (OPPOSITE); KITT PEAK NATIONAL OBSERVATORY (ABOVE)

Suggesting a double loop of luminous beads, the sequence of images opposite traces the ten-billion-year life cycle of our sun—typical of a kind and size of star very common in the Milky Way galaxy. Born in a condensing interstellar gas cloud much too large to show in true scale, the new star begins to shine by its own nuclear process, and quickly matures to the long-lasting yellow stage. Near the end of its lifetime (still five billion years away for our middle-aged sun), the star expands to become a red giant, searing any planets within its fiery envelope. After a brief period of contraction (not pictured), the star swells again to red-giant size. With most of its nuclear fuel expended, the core of the star

remains as a white dwarf and the outer layers dissipate into space. Finally the white dwarf cools to a black-dwarf cinder.

Turbulent birthplace of stars, the Lagoon Nebula (above) glows with the red light of hydrogen. Ultraviolet radiation from bright, newly formed stars embedded in the cloud heats and drives away the interstellar gas. Deep within the cloud, small dark globules of obscuring matter emit strong infrared radiation, probably indicating the beginnings of star formation not visible to the human eye. These protostars have not yet reached the critical stage of density that will ignite their internal nuclear reactors and cause them to shine.

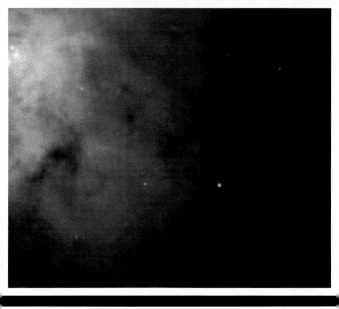

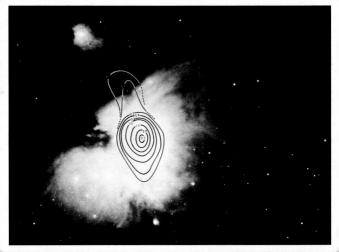

Studded with bright young stars, the Great Nebula in Orion (left) pro
vides a relatively nearby object for investigation of stellar birth. A
small section of the Great Nebula (above, upper) illustrates the Inter
active Picture Processing System developed at Kitt Peak. The IPPS
uses a computer and videotape to process data collected from a
faint object; astronomers can then see an improved image of the
object on a television screen.

A new field of astrochemistry began in 1963 with the discovery of
radio emission from interstellar molecules. A contour map super
imposed on a photograph of the Orion nebula (above, lower) indi
cates varying formaldehyde intensities in its dense molecular cloud.
Molecular studies can yield information about temperature, density
mass, motion, and radiation fields within optically opaque clouds sur
rounding regions of star formation. Some scientists believe planets
may form from molecular clouds along with the central stars.

Billowing clouds of gas and dust veil the stellar nursery of the Orion nebula, 1,500 light-years from earth. This artist's view portrays a region containing stars in several stages of formation.

In the center of the nebula, the brilliant stars called the Trapezium illuminate the turbulent clouds with blue light. Stellar ultraviolet radiation creates the green light of ionized oxygen, and the red of hydrogen and ionized sulphur and nitrogen. Radiation pressure from the Trapezium group has hollowed a cave in the dark cloud.

At the upper left of the painting, a young star shines with hot blue light. Some newborn stars, deeply embedded in clouds, have their light absorbed and transformed to heat by the surrounding cocoons of dust and gases. Astronomers must use infrared detectors to study these; one glows dully to the right of the bright young star, and one below. Red light from a proto-star—a body still condensing—bursts through a layer of cloud to spotlight another layer, above and to the right of the Trapezium.

Below the red protostar, an older star colliding with interstellar gases creates a white bow wave and wake much like those formed by a boat plowing through water.

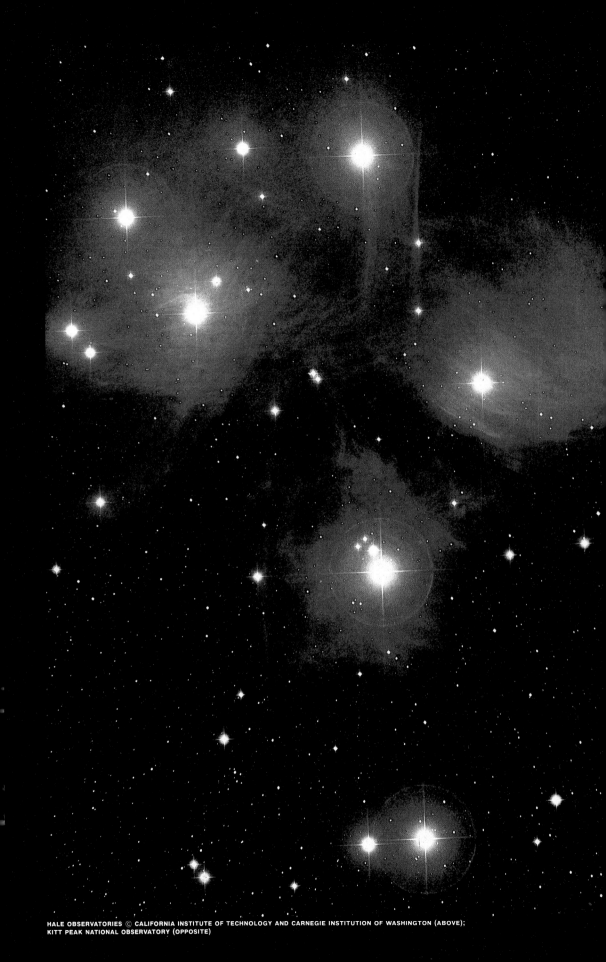

Known from antiquity as the Seven Sisters, the Pleiades star cluster (opposite) in the constellation Taurus offers a glimpse of a more advanced stage in stellar evolution. A familiar object in the sky of the northern hemisphere, the Pleiades appears to the unaided eye as a small cloud of light, with perhaps six to ten of its brightest stars distinguishable on a clear dark night. Actually this field includes about 300 stars that have formed over the last 70 million years since a dense cloud of gas and dust began to collapse under gravitational force, breaking into many smaller masses, or protostars. The photograph shows that the brightest stars in the cluster lie in a haze that scatters their hot blue light and appears as multiple filaments.

Because the human eye lacks color sensitivity at low levels of light, we see the blue filaments as white without the aid of color film. A longer exposure (above) depicts even more clearly the streaks of interstellar material, probably left over after the stars condensed from the cloud.

must fire the stars. Physicists of the time ridiculed his idea that a star's interior could be hot enough for nuclear fusion. In exasperation, Sir Arthur responded: "The critics lay themselves open to an obvious retort; we tell them to go and find a *hotter place*."

Today the details of nuclear fusion in stars are well understood by scientists. When four hydrogen atoms combine, or fuse, to form one helium atom, a small amount of mass is lost in the form of radiated energy. Each second, the sun converts 564 million tons of hydrogen to 560 million tons of helium. The four-million-ton difference in mass is transformed to energy in accordance with Einstein's famous formula $E=mc^2$ (energy equals mass multiplied by the square of the speed of light). In simple terms, the formula means that small amounts of mass can be transformed into tremendous bursts of energy. In fact, fusion is ten million times as efficient as chemical burning—of, say, oil or coal—in converting matter to energy.

I N SPECTRAL TYPE the sun is a class G star, midway between the hottest blue stars and the coolest red stars. From earth it appears a hundred billion times brighter than any other star because it is so close to us, but it would appear relatively feeble if matched against giant stars at the same distances. Rigel, a blue supergiant in the constellation Orion, is 15,000 times brighter than the sun. At the other extreme, the white dwarfs are 20,000 times dimmer. In size, the sun also stands in mid-range. Thirty million suns could fit within Antares, a red supergiant, but white dwarfs are so small that a hundred thousand could be packed into a sphere the size of the sun.

Although there are vast differences in stellar sizes and brightness, most stars fit somewhere in a continuous sequence ranging from brilliant, hot blue stars of larger mass to dim, cool red stars of smaller

mass. When brightness is plotted against color or temperature, as you will see in the diagram that appears on page 88, these "main sequence" stars lie in a narrow diagonal belt. The brightness varies by a hundred thousandfold from one end of the stellar belt to the other, but the variation of mass and size for these stars is relatively small. Stars that have completed most of their life cycle—brilliant giants and extremely faint dwarfs—have evolved away from the main sequence.

Evolution is rapid for the heaviest stars, which burn their energy at a spendthrift rate and leave the main sequence after only a million to ten million years. Middle-aged, medium-sized stars like our sun have spent billions of years on the main sequence, but will move appreciably off the track sometime during the next several billion years. For the smallest stars, evolution off the main sequence may take more than a hundred billion years.

With computers we can trace the evolution of typical stars much, much faster than was possible even a generation ago. The magnitude of such computations is so great that old-fashioned hand calculators would take thousands of man-years to complete the evolutionary story of one typical star. The fastest computers now available can perform the same mathematical operations in minutes.

Here in summary is the pattern our sun and similar stars are likely to follow as they evolve away from the main sequence:

Four or five billion years from now, when fusion has converted about an eighth of the solar core to helium, the sun's metabolism will begin to show signs of serious upset. Because the temperature is not high enough to ignite helium, the core will contract under the pull of gravity. The temperature in the center will rise steadily, hydrogen outside the core will ignite, and the region of nuclear fusion will move from the core to the surrounding hydrogen shell. Continuing compression of the core will raise its temperature still higher; the surrounding hydrogen will get hotter, and the rate of nuclear reaction in the hydrogen shell will increase. It may seem a paradox, but extinction of the nuclear fire at the very center of the star will actually lead to a greater outpouring of nuclear energy from outside the core.

As the increasing heat reaches the outer layers, they will begin to swell. More and more heat from the burning hydrogen shell will force trillions of tons of overlying hydrogen to expand outward hundreds of thousands of miles. While the star is growing larger, its surface temperature will drop perhaps 2,000°, giving it a reddish hue. By the time it is nine to ten billion years old, it will be about twice the diameter, and twice as bright as it was when its nuclear furnace first began burning.

From then on, the pace of evolution accelerates. In the next billion years the star doubles again in diameter, ballooning toward red giant size. Then, in only a hundred million years, it bloats 50 times larger and shines 500 times brighter.

As it expands on this gargantuan scale, the sun will scorch its four inner planets. Earth will be cooked with hundreds of times the present flood of solar heat. Oceans will boil away, and the vapors will cloak the earth in thick clouds.

The next stage is complex but short-lived. Fusion of helium atoms begins, but is quickly interrupted by a core explosion —the "helium flash." After the flash, the core temperature drops, helium fusion stops, hydrogen-shell burning slows down, and the swollen envelope starts to contract. About 10,000 years pass as the star shrinks and its brightness fades.

Then the central temperature rises again to more than a hundred million degrees, and helium begins to change into carbon. When all the core helium is converted to carbon "ash," the zone of helium-burning moves outward. Back toward the red giant phase zooms the star, about a hundred times faster than it took to get there the first time. (Continued on page 91)

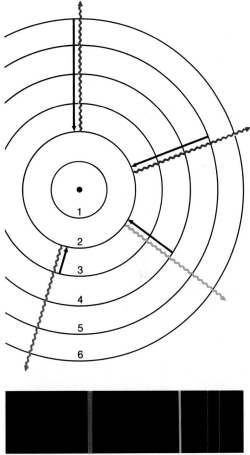

GEOGRAPHIC ART

Fingerprint of the hydrogen atom appears as a specific spacing of different wavelengths (on black above) in the spectrum's visible-light area. The stylized diagram (top) shows the inside of the atom and how it emits light. Hydrogen's single electron circles the nucleus (black dot) in distinct orbits, each corresponding to a specific level of energy. Collisions with other atoms give the electron extra energy to jump to a wider orbit. When the electron falls back (black arrows) to a smaller orbit, it radiates energy as light (wavy lines correspond to lines seen in the spectrum). The electron's fall back from bigger orbits produces different wavelengths: from 3rd to 2nd, red light; from 4th to 2nd, blue; from 5th, violet; from 6th, purple. Red represents the least amount of energy, purple the most.

JOHN LUTNES, KITT PEAK NATIONAL OBSERVATORY (OPPOSITE); WILLIAM L. ALLEN, NATIONAL GEOGRAPHIC STAFF (TOP); J. R. EYERMAN

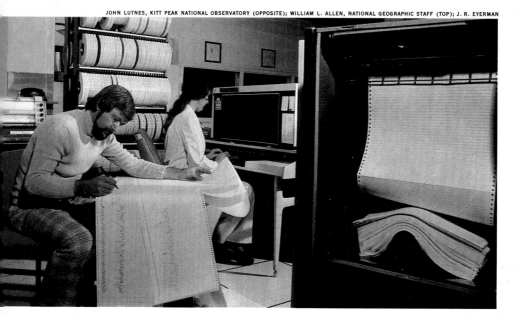

Complex instruments and ingenious techniques play key roles in star-watching at Kitt Peak. A diffraction grating (opposite) breaks up light into the bright bands of the spectrum. Starlight gathered by a telescope disperses off the fine parallel grooves of the grating into different wavelengths of light; a magnetic tape records the intensity of each wavelength. In the computer room (above), the data goes into a plotter. Don Hall examines the end product — a highly accurate graph of a star's spectrum. Computer-processed photographs reveal surface features of Betelgeuse, a red supergiant (below). In 1974 S. Peter Worden, Roger Lynds, and John W. Harvey (left to right) used the Mayall Telescope at Kitt Peak to take multiple exposures of the star. They then used a computer to scan the images and reject the effects of atmospheric distortion. Here, with Don Wells (far right), they produce a picture of the surface of Betelgeuse — the first image of the surface of a star other than our sun.

SPECTRAL   CLASS

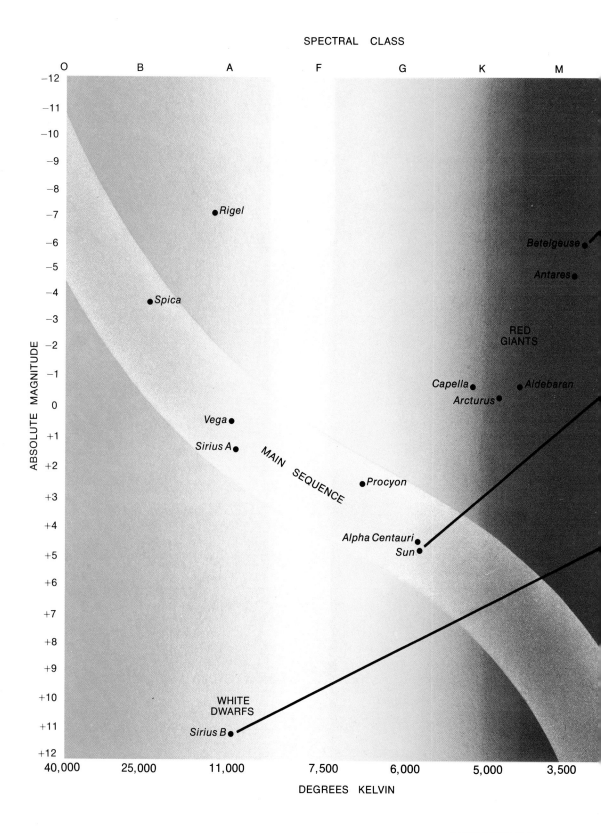

ABSOLUTE MAGNITUDE

DEGREES   KELVIN

H-R DIAGRAM NAMED FOR ITS DEVELOPERS,
EJNAR HERTZSPRUNG AND HENRY RUSSELL; GEOGRAPHIC ART

Plotting the Stars

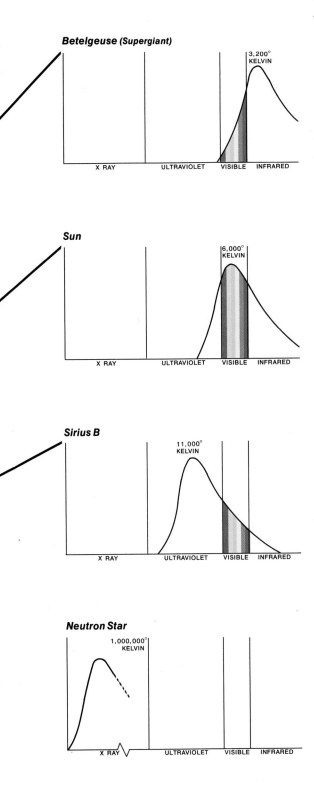

Betelgeuse (Supergiant)

3,200°
KELVIN

X RAY ULTRAVIOLET VISIBLE INFRARED

Sun

6,000°
KELVIN

X RAY ULTRAVIOLET VISIBLE INFRARED

Sirius B

11,000°
KELVIN

X RAY ULTRAVIOLET VISIBLE INFRARED

Neutron Star

1,000,000°
KELVIN

X RAY ULTRAVIOLET VISIBLE INFRARED

H-R diagram shows where stars appear when scientists plot them on a chart to study the relationship between brightness and temperature. Letters across the top refer to the spectral class, ranking stars according to their relative temperatures, with a class O blue star the hottest and a class M red star the coolest. The figures across the bottom of the diagram show actual surface temperatures. On the left side of the scale, figures indicate absolute magnitude: a measure of a star's luminosity, or actual brightness. A star of −12 absolute magnitude would be among the brightest. Another term, apparent magnitude, refers to the brightness of a star as observed from earth. Since distance modifies brightness, astronomers measure a star's apparent magnitude, then calculate its absolute magnitude by reference to the distance, measured independently. A person with good eyesight could see a star in the night sky with an apparent magnitude of +6. Any star below that point on the scale would reveal its presence only to telescopes. Most stars lie along the curved diagonal band known as the main sequence and range from hot blue stars to cool red stars. There they spend the greater part of their lives burning hydrogen. Our sun, for example, a star of average size and middle age, will stay on the main sequence another five billion years. A class G star, its surface temperature measures about 6,000° Kelvin (5,700° C.).

Schematic drawings indicate the predominant radiation emitted by four typical stars with different surface temperatures. Betelgeuse emits mostly infrared rays. Our sun peaks in the visible light portion and also emits invisible ultraviolet and infrared. Sirius B radiates most of its energy as ultraviolet rays. Extremely hot, a neutron star emits almost all its radiation as X rays. The broken line at the base of the spectrum and the dotted lines in the spectral curve indicate that the X-ray portion extends far to the left of the drawing and does not appear in true scale.

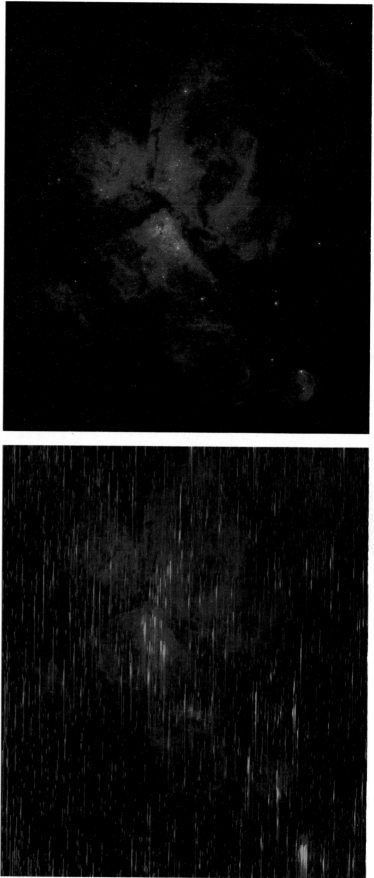

Nursery of new stars, the Carina nebula contains some of the bluest, most massive stars in our galaxy. Easily visible to the unaided eye, the nebula presents a large region of ionized hydrogen and dust clouds in the southern portion of the Milky Way.

A mysterious newcomer, Eta Carinae (near the center of the nebula at right) shone as an unimpressive star until 1843. Then it blazed and rivaled Sirius as the brightest star in the sky; scientists believe it erupted as a nova. After 15 years the star began to dim, and by 1867 had vanished from view except through telescopes. Over the years, a small nebula slowly developed around Eta Carinae. Interest in both the star and its small nebula revived in the late 1960's when Gerry Neugebauer and James Westphal of the California Institute of Technology pinpointed Eta Carinae as one of the strongest sources of infrared radiation.

Stellar rainbows streak another photograph of the Carina nebula. A large prism placed before a telescope breaks the light from its stars into a spectral array. Astronomers working with similar streaks, or spectra, can classify stars, determine their temperature, and then plot them on the H-R diagram (pages 88-89).

CERRO TOLOLO INTER-AMERICAN OBSERVATORY

Explosive flashes in the helium shell may cause the star to fluctuate in size several times as it moves toward maximum expansion.

If the mass is relatively small, the second approach to the red giant stage is the star's last gasp. The carbon core never gets hot enough to fuse; but as its temperature rises, the outer envelope expands and cools. The surface layers of the bloated envelope are so loosely held by gravity that the radiation pressure of light, generated near the surface, is sufficient to blow the outer layer off the star. A vast, tenuous shell of gas balloons far out into space. As this expands and thins out, the core left behind becomes visible as a white dwarf star. Its ultraviolet radiation excites the expanding gas shell to display the beautiful fluorescent colors of a planetary nebula.

The white dwarf core of carbon continues to contract until it has a radius of a few thousand miles and a density of some tens of tons per cubic inch. Because the star is now as small as earth, it appears very faint even though its surface temperature is higher than that of the sun.

This late stage of the cycle sometimes produces a brilliantly flaring star of a kind we see two or three times a year and call a nova, after the Latin word for new. As it shrinks toward invisibility, the white dwarf may suddenly flash hundreds to millions of times brighter, giving the illusion of the birth of a new star. What triggers the outburst is not clearly understood. One clue is that many novae belong to binary, or double, stars; perhaps, in a binary system, a sudden flood of matter falls onto a white dwarf from its companion star and provides fresh fuel for the short-lived flare.

A nova begins to fade almost at once, rapidly at first, then more slowly; typically the star has returned to its earlier faint level within a few months to a year.

As they age, white dwarf stars gradually cool from white to yellow to red; finally, after trillions of years, they become cold black cinders in the stellar graveyard.

When the sun finally collapses to a white dwarf, the earth too will cool. As the rain clouds empty, oceans again will fill their basins, and carbon dioxide snow will fall. Increasing cold will grip the earth, the oceans will freeze, and an everlasting ice age will settle over the planet.

EVEN THOUGH they were already aware of great variation in the kinds and sizes of stars, astronomers were nonetheless startled when they first realized the nature of white dwarf stars. They knew that stars comparable to the sun have an average density equal to that of water and are gaseous to the core; now they were confronted with stellar material harder than steel, and ten thousand times as dense!

Sirius, the brightest star in the sky, had been observed and described since ancient times, but not until 1862 did astronomers actually see that Sirius is two stars. The presence of the companion in the binary system had been detected earlier; it circles about the bright star every 50 years, tugging it back and forth across the line of sight in a gravitational embrace.

The somewhat wagging motion of Sirius, as well as its distance, was carefully measured, and from these figures the masses were computed. Sirius A, a very luminous white star, turned out to be about 2.5 times as massive as the sun. The mass of Sirius B, the faint companion, was almost the same as that of the sun, but its light was 400 times weaker. Those figures were not especially surprising, for thousands of faint red dwarf stars were already known.

But when in 1915 Walter Adams at Mount Wilson found that the companion star was not a red dwarf but a white one, astronomers were incredulous. Some arithmetic helps us see why:

The fact that the star was white implied that it was hotter than the sun, and roughly three times brighter per square mile of surface; yet the star was 400 times *less* luminous than the sun. Accordingly, its total

High on Kitt Peak in Arizona, the Mayall Telescope (opposite) gleams softly against the night sky. Rotation of the earth during a 30-minute exposure streaked the photograph with the light of stars of various colors as they described arcs around bright Polaris, the North Star. Inside the dome (right), a tiny window in the telescope reveals the face of an observer in the prime-focus cage. In the control room (below), Hal Halbedel positions the 300-ton telescope. Once the astronomers point the computer-controlled telescope at a particular object, it automatically tracks the target. Red lights and blackout curtains help astronomers maintain their night vision.

N.G.S. PHOTOGRAPHER JAMES P. BLAIR (TOP); DAVID L. MOORE, KITT PEAK NATIONAL OBSERVATORY (OPPOSITE); WILLIAM L. ALLEN, NATIONAL GEOGRAPHIC STAFF, AND WARREN HILL, KITT PEAK NATIONAL OBSERVATORY

radiant surface must be 1,200 times *smaller* than the sun's. That translates to a diameter 35 times smaller and a volume 40,000 times less. Imagine a mass equal to that of the sun packed into such a small space! As Eddington put it: "The message of the companion of Sirius when it was decoded ran: 'I am composed of material 3,000 times denser than anything you have come across; a ton of my material would be a little nugget that you could put in a matchbox.' What reply can one make to such a message? The reply which most of us made in 1914 was—'Shut up. Don't talk nonsense.'"

But while astronomers searched for errors, more white dwarfs were found. Then within a few years Eddington provided the explanation: At the extremely high temperatures in the core of the white dwarf, the atomic nuclei are stripped of electrons and then can be crushed to a density not merely thousands but billions of times that of gold.

Although the eye cannot penetrate the surface of a star, Eddington's theory could probe directly to its heart. From the base provided by his theoretical models, and with the aid of computers, astrophysicists have become actuarial experts on the society of stars, accounting not only for the common, average specimens but also for the freak dwarfs and giants.

Still, none of us was quite prepared for X-ray stars, those rare celestial bodies whose predominant emission of energy is in the form of X rays. Their discovery was as much a shock as the first white dwarf.

In 1962, American rocket astronomers Riccardo Giacconi, Herbert Gursky, Frank Paolini, and Bruno Rossi made the surprising discovery that X rays were coming from a region of sky near the center of the galaxy, in the constellation Scorpius. Shortly afterward, my colleagues and I zeroed in more precisely on the X-ray emission. Our rocket instruments identified an X-ray source that we named Scorpius X-1, blazing like a hundred-million-degree inferno against a visible backdrop of the most commonplace stars. When a visual match-up was achieved a few years later by Giacconi, working with optical astronomers, Scorpius X-1 was identified as a faint blue star of innocent appearance. The mystery of this astonishing X-ray powerhouse, so well concealed from optical telescopes, provides a continuing challenge for today's Eddingtons.

THE STARS THAT PROVIDE the heavens' most dramatic spectacle are the rare supernovae —stars that explode in a cataclysmic event, releasing 10,000 to a million times more energy than a nova eruption. Stars less than about 1.4 times as massive as the sun can never explode, because their temperature cannot climb high enough to ignite carbon, and the fusion process stops with hydrogen and helium. Such stars live out their declining years as fading white dwarfs, and the smaller they are the longer they last.

For heavy stars, however, it is a different story: The bigger they are, the harder they fall. Evolution is relatively rapid, and for some the end is violent collapse of the core accompanied by a spectacular explosion of the outer layers. It is this process that produces two fascinating phenomena: the supernova and the neutron star.

If a star begins its stay on the main sequence with a mass greater than about ten times that of the sun, fusion doesn't end with the build-up of a carbon core. Instead, the excessive weight relentlessly forces shrinkage that heats the carbon above 600 million degrees and initiates fusion to oxygen, neon, and magnesium. When the carbon is exhausted, the core again shrinks and heats up, and higher-temperature fusion in successive steps transforms magnesium to sulphur and, finally, sulphur to iron. By this time the core is surrounded by shells of lighter elements, with hydrogen on the outside.

The periods between successive collapses can range from 10,000 years down to less than 100 years in the later stages of fusion. When the star has evolved its iron core, death is imminent. As the temperature creeps above five billion degrees, high energy particles called neutrinos are produced in greater and greater numbers. Moving at the speed of light, they flood out of the star fast enough to drain the energy of the core in a single day. Only a faster and faster collapse can sustain the high temperature which, in turn, speeds up the avalanche of neutrinos, until collapse becomes a runaway process.

At about six billion degrees, there is an abrupt switch in the trend of nuclear reactions. Iron begins to disintegrate into helium and neutrons. The star must now draw on its gravitational reserve to pay back all the energy it had previously bought by fusing elements. The price is complete collapse; in only a second or two, the star caves in with a devastating crash. All the mass piles up in the center at a density of a billion tons per cubic inch. Electrons are squeezed into protons to form neutrons, and the entire core becomes a supernucleus of a billion trillion trillion trillion trillion neutrons — an incredibly dense neutron star, possibly about ten miles in diameter!

Yet, in its final throes the dying star manages one last flash of splendor: the supernova. As the core collapses, the cooler external layers of the star suddenly lose their support; the outer material rains down freely, and its dynamic energy is rapidly transformed to heat. The temperature shoots up to accelerate nuclear burning of hydrogen and the lighter elements in the outer shell. Above a billion degrees, in little more than a second, oxygen near the surface is consumed completely in a colossal explosion.

If the explosion's debris contains as much as one solar mass, the supernova may shine as brightly as a full galaxy of stars for several seconds, and may give as much light as 200 million suns for two or three weeks. From the explosion a brilliant nebula emerges around the neutron star, the burned-out core of the original star.

I must emphasize that all this is purely

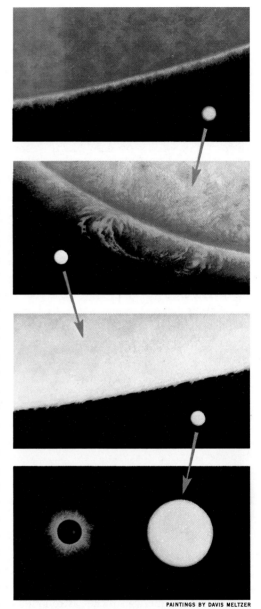

PAINTINGS BY DAVIS MELTZER

From giant star to black hole: A red giant (top panel) looms in contrast to our sun, a medium-size star. The sun, in turn, appears huge beside a white dwarf, and the white dwarf seems vast next to a neutron star about ten miles wide. Even the neutron star (bottom) looks large beside a black hole, hypothetical result of the runaway collapse of a very massive star. Not even light can escape the crushing gravity of the tiny black hole.

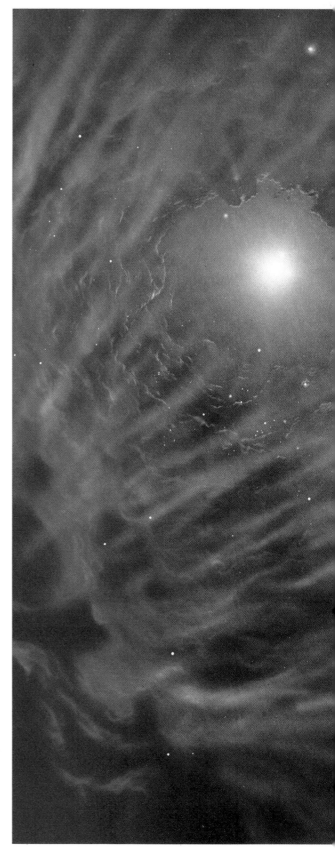

Stellar Shock Wave

Shock wave in the Orion nebula arcs around a large star (at right). Although the wave looks much like a blue wake produced by a spaceship zooming through an interstellar sea of green, the reverse seems true. Astronomer Ted Gull, who discovered the phenomenon in 1974, believes the interaction of ionized gas flowing past the star produces the shock wave.

At left, blue-white stars of the Trapezium group gleam within a celestial cavern carved out of a dark cloud by their own intense ultraviolet radiation. The rippling red clouds, at the surface of the dark cloud, contain neutral and ionized gas that radiates the colors of excited hydrogen and ionized oxygen and nitrogen. The expanding ionized material flows rapidly outward to the less dense area beyond the star. The hot gas, characterized by the brilliant green emission of ionized oxygen, surges past the star at speeds faster than sound, creating the shock wave.

theoretical, a good fit to what we have observed in the heavens. Other models have been proposed, but in any scenario the time scale for the entire collapse and explosion can hardly exceed a few seconds. The star, which required several million years to form, and then may have lived tens of millions of years in various states of equilibrium, hurls much of its matter back into interstellar space in one brilliant flash.

Although the theory had been worked out in detail, no observed star had yet been interpreted as a neutron star when in 1963 our rocket instruments pinpointed Scorpius X-1 and, on the same mission, detected X-ray emission from the Crab Nebula, a supernova remnant. A short time later I described our measurements at the Institute for Advanced Study at Princeton, New Jersey. As I began my lecture, the famed physicist J. Robert Oppenheimer entered the room and sat down immediately in front of me. Very tentatively, I pointed out the possibility that X rays from the Crab might come from a neutron star, the dense core left over from the great explosion that produced the nebula.

That evening a group of us, including Oppenheimer and a young astronomer, Donald Morton, were dinner guests of the Danish astrophysicist Bengt Strömgren. Morton had done research on stellar structure and was able to confirm that a neutron star could be hot enough to radiate X rays. Our conversation set Oppenheimer to reminiscing about how he and his colleagues in the late 1930's had theorized the formation of neutron stars, but had no real hope that these incredibly small stars would ever be observed — certainly not visibly, and who would seriously have anticipated X-ray astronomy?

His excitement was obvious. World War II had diverted him to Los Alamos and the atomic bomb project, and neutron stars had long been relegated to the back of his mind. Almost a quarter century had passed, and now the strangest of theoretical stars might be about to materialize!

Since the concepts of neutron stars and supernovae are interrelated, let's further explore these rare and spectacular celestial explosions. The rate of occurrence of supernovae in the galaxy is believed to average about one per century, yet in the last thousand years only three have been seen. Perhaps others have occurred behind obscuring dust clouds. It is remarkable that the last two occurred within one generation and were observed by the two greatest astronomers of their time, Tycho Brahe in 1572 and Johannes Kepler in 1604.

By all odds we are overdue for another, and what a glorious gift that would be to modern astronomy! But probabilities, of course, offer no precise guarantees.

SHORTLY BEFORE DAWN on July 4, 1054, the Chinese court astronomer Yang Wei-Te noticed a brilliant star in the eastern sky over Khaifeng, capital of the Sung Dynasty. The startled Yang realized at once that the star had not been there before. Chinese astronomers were very precise about such matters; over a period of nearly 1,300 years, for example, they had recorded all 17 appearances of a comet which later came to be named for Edmund Halley. Because the star was reportedly yellow, the imperial color, the emperor ordered Yang to make a prediction as to its significance. Since the "guest star," Yang responded, did not infringe upon the red giant star Aldebaran, it was taken as a portent that the country would attain great power; and the event was duly noted in court records.

The guest star shone as brightly as the planet Venus and was easily visible in daylight for the next 23 days. Only after 650 nights did the star finally fade from view.

The position given in the Chinese record is near Zeta Tauri. A similar report appears in Japanese chronicles. Although the star must have been widely seen in Europe and the Middle East as well, no account exists. But in the early 1950's, William

C. Miller of the Hale Observatories found two prehistoric Indian rock drawings in northern Arizona showing what is believed to be the supernova. The star appears near a crescent moon, a juxtaposition that would have been visible on July 5, 1054. Since then, nine other similar rock art examples have been found.

Today the spectacular nebula associated with the supernova of 1054 is one of the most beautiful, most significant, and most constantly studied objects in the sky. It was first discovered by the Englishman John Bevis in 1731. It appeared at the tip of the horn of the bull in the constellation Taurus. Charles Messier, a French astronomer whom Louis XIV called the "comet ferret," found it independently in 1758; it is listed in his catalogue of nebulae as M 1. In 1845 Lord Rosse first noticed the filamentary structure and gave it the name Crab Nebula.

Measurements made by John C. Duncan at Mount Wilson Observatory in 1921 showed that the nebula was expanding. Later, astronomers keyed its present size to its origin almost 900 years earlier. The intricate beauty of the Crab began to emerge in Walter Baade's photographs taken at Mount Wilson in 1942. We now see the Crab Nebula as a dazzling web of bright red filaments enveloping a diffuse cloud of bluish white.

The debris of the Crab Nebula amounts to about one-tenth the mass of the sun and is expanding at about 700 miles per second. In the span of more than nine centuries its diameter has grown to about six light-years. At the center there appear two faint stars, one of which is now known to be a neutron star, the collapsed core of the star that exploded. The debris is moving fast enough to escape from the neutron star, and will eventually dissipate into space, enriching the interstellar medium and providing material for the formation of new stars.

Although the filaments of the Crab Nebula produce spectra that are typical of expanding ionized gases, the white light of the diffuse part of the nebula is unique, and

ROBERT M. LIGHTFOOT III (ABOVE); HERMAN ECKELMANN, CORNELL UNIVERSITY

Studying the evolution of stars, S. Chandrasekhar (above) showed theoretically in 1931 that a star less than 1.4 times the mass of the sun must end its life as a white dwarf. This mass—the Chandrasekhar Limit—provides a dividing line between stars that become white dwarfs and larger stars that may end as incredibly dense neutron stars or black holes. Thomas Gold (below) explained the strange pulsating radio source in the Crab Nebula as a spinning neutron star: Its rapid spin provides the energy for its radiation.

cannot be produced by any atomic emission process. If it came from a moderately hot, rarefied gas, such as that found in most glowing nebulae, we would expect characteristic spectrum lines of hydrogen, helium, and oxygen; but no emission lines appear. A solution to this puzzle was proposed by the Soviet astrophysicist Josef Shklovsky: that the light comes not from ionized atoms, but from free electrons spiraling at high speeds through a magnetic field. Nuclear physicists call such light "synchrotron radiation," because they have produced it in high-energy machines called synchrotron accelerators. It is highly polarized — that is, the light rays' vibrations tend to be aligned in a plane rather than randomly oriented, so that certain features appear and disappear as a polarizing filter is rotated.

Shklovsky deduced that the electrons in the Crab move at nearly the speed of light in magnetic fields about a thousandth as strong as that of the earth. Since electrons of such energies are found in cosmic rays, he also proposed that galactic cosmic rays might be produced primarily in supernova remnants like the Crab; but it was difficult to understand how they could be present with such high energies almost a thousand years after the explosion. Could the remains of the stellar core still be pumping high-energy electrons into the nebula?

The answer is yes; but this only became apparent with the discovery that the core is a neutron star, and also a pulsar, or rapidly pulsing source of radiation. Pulsars will be discussed further in the next chapter.

THE SYNCHROTRON RADIATION from fast electrons could, theoretically, extend from gamma rays and X rays to the longest radio waves. Furthermore, if the residual core of the supernova had formed a neutron star, the predicted surface temperature would lie in the vicinity of ten million degrees, sufficient to make the star itself a strong thermal X-ray source. We were excited to find, in the earliest X-ray observations attempted with rockets, that the Crab radiates X rays with ten thousand times as much power as the sun produces in heat and light. But the X-ray detectors of that era had a wide field of view and could not tell the difference between emissions from a broad nebular source and those from a specific starlike point.

About every nine years, the moon passes directly between us and the Crab Nebula. Early in 1964, I was looking over predictions of when the moon would block out radio waves from the heavens — a list prepared for our astronomers by the Naval Observatory — when I noticed that an eclipse of the Crab Nebula would occur on July 7. The eclipse offered the possibility that we might distinguish a point X-ray source embedded in the broad nebulous emission. Even though the X-ray detector took in the entire nebula, the moon passed between us and the Crab so slowly that it could serve as a sharp-edged shutter on the radiation. Suppose the X rays came from a neutron star; they would cease abruptly when the edge of the moon intercepted the beam. If, on the other hand, the X rays were produced over the full nebula, as the visible synchrotron radiation is, they would grow gradually but steadily weaker as the moon sliced across the Crab and eclipsed the nebula.

We promptly set about readying a rocket payload for the attempt, although time was short and the schedule very tight. While we were making our preparations, I received an unanticipated stimulus: A letter arrived from Josef Shklovsky urging me to attempt such an observation. He explained that the Soviet rocket astronomy program was not yet ready for such an experiment, but he was sure that we could bring it off. He considered the mission highly important for two reasons: the possibility of detecting a neutron star, and of measuring the size of the X-ray nebula.

Whatever uneasiness we may have had about the difficult project was dispelled by

Shklovsky's letter, and we pushed ahead with greater determination to meet the target date. In 1964, however, the mechanics of rocket astronomy was still rather primitive. A gas jet stabilizer was being developed to point the Aerobee rocket with an acceptable accuracy, but it had failed some half-dozen tests in two years. Furthermore, the flight time of the rocket was only about four minutes, and the eclipse would last about 12 minutes. It was essential to launch within seconds of the eclipse maximum and to achieve a highly accurate trajectory. The odds seemed heavy against success, but the stakes were worth the gamble.

On July 7 the rocket flew exactly as planned, and the stabilizer worked well for the first time. The results of our observation, though valuable, were less completely satisfying. Because the X rays showed only a gradual decline with the passage of the moon, we were forced to conclude that they came from the broad nebula. If the neutron star existed, it must have cooled down enough that its thermal X rays were too weak for us to detect. We had to be content with the knowledge that our observations had identified the X-ray source with the synchrotron radiation of the nebula.

The incident well illustrates the building-block process of scientific research. Actually the evidence for a neutron star was concealed in very rapid X-ray pulsation; and since astronomers did not yet know about pulsars in 1964, our instruments were not designed to detect intermittent X-ray signals. Only three years later, the discovery of radio pulsars set the stage for a series of observations that proved the existence of a central neutron star in the Crab.

The violent ejection of matter that accompanies a supernova explosion usually drives a shell of gas outward through interstellar space. As the shock wave expands, dust gathers in a thin layer at the shock front.

Each supernova produces its own expanding bubble. Over the ages supernovae have made the *(Continued on page 108)*

Nova Herculis 1934 flares to near-maximum brightness (above, upper) on March 10, 1935. By May 6 (above, lower), the star had faded to normal brightness. Although scientists estimate that 40 novae occur in our galaxy each year, they detect only two or three. In a spiral galaxy (below), a supernova erupts in 1959, and then fades. A supernova results from an explosion much more catastrophic than the eruption of a nova. The last visible supernova in the Milky Way, recorded by Johannes Kepler, occurred in 1604.

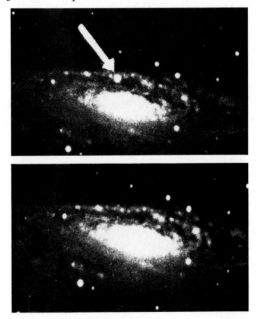

The Crab Nebula

SCALE: SEE PAGE 4

Reddish filaments fringe the Crab Nebula, a rapidly expanding web of gas left over from the explosion of a dying star. Chinese chronicles recorded the supernova as a "guest star" that appeared on July 4, 1054, and shone so brightly that it remained visible in daylight for 23 days. An Indian pictograph in northern Arizona (above) may portray the same event. William C. Miller, former chief photographer for California's Hale Observatories, found the pictograph in 1952 while following his hobby of archeology "as a way of getting away from astronomy." Now some scientists believe at least nine other examples of rock art found in the Southwest, and another in Baja California, also record the exploding star. Calculations from lunar tables indicate that Indians could have seen the supernova near a crescent moon about the time the Chinese saw the guest star.

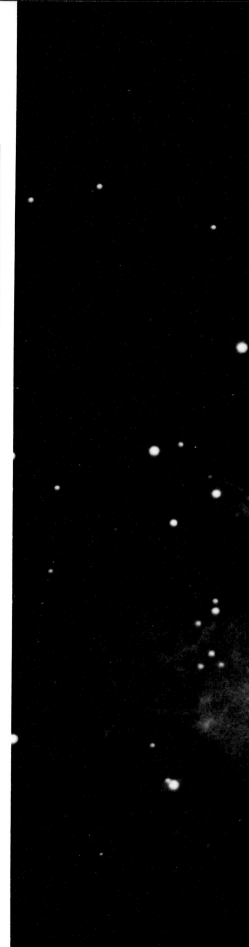

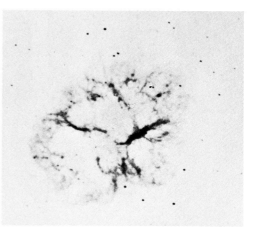

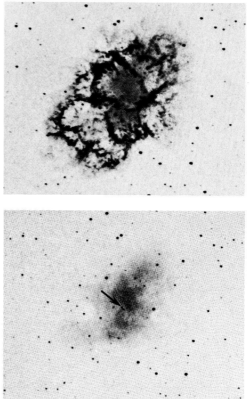

The Crab Nebula

Former director of Kitt Peak National Observatory, Nicholas Mayall (above, left) makes some of the first observations with the telescope named in his honor. Work by Dr. Mayall and others led to conclusive identification of the Crab Nebula as the remnant of the exploding star of 1054. Physicist William A. Fowler (left) uses an electrostatic accelerator to study nuclear processes that occurred in formation of the Crab's neutron star. High-energy electrons in the Crab account for its radio, optical, and X-ray radiation.

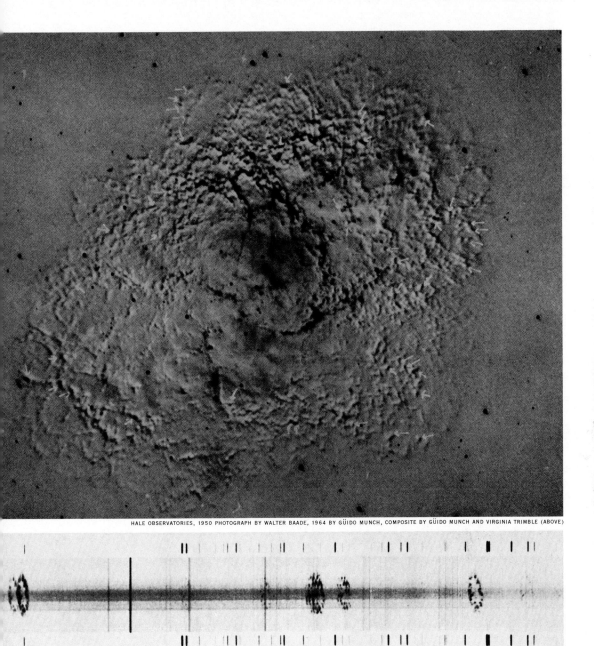

Images of the Crab Nebula, taken at several wavelengths, reveal widely differing structures. The top photograph (opposite) of ionized oxygen (dark areas) in the Crab shows the interaction of debris from the explosion of the star with interstellar gas. In the middle photograph, deeper inside the Crab, ionized sulphur appears. The bottom photograph shows the light of synchrotron radiation in the heart of the nebula. The arrow indicates the Crab's pulsar, the core of the exploded star and a source of powerful radiation. Spinning like a runaway lighthouse, it emits a beam of radiation that reaches earth in pulses at the rate of 30 times per second.

Two other photographs, taken 14 years apart and superimposed (top), show expansion of the Crab Nebula. In the 1950 version, each gas filament appears white. In 1964, the filaments (printed dark for contrast) lie outside the white filaments, thus indicating expansion. The spectrogram (band above) shows optical radiation from a slice across the nebula. The gas filaments, within an expanding outer shell, appear as ragged ovals. The light of synchrotron radiation streaks the length of the spectrogram.

Part of a vast supernova remnant called the Cygnus Loop, the Veil Nebula arcs across the constellation Cygnus. Roughly 50,000 years old, the remnant continues to expand

in an ever-widening circle. Some 25 of these filamentary loops appear in the Milky Way; they represent the borders of supernovae debris much older than that of the Crab Nebula.

SCALE: SEE PAGE 4

interstellar medium look like a cosmic bubble bath, a dusty froth that provides the observed pattern of dust lanes in our galaxy. About 25 such shells have been identified in the Milky Way. Away from the bright nebulae, these dust clouds show a fine filamentary structure intertwined like lace.

If a supernova remnant is not too old, the shock-heated gas at the bubble's surface may still be hot enough to radiate X rays. An example is the Cygnus Loop, about 50,000 years old. The older the bubble, the lower the surface temperature and the weaker the X-ray emission. Thread-like loops of dust even extend well above the plane of the Milky Way and may explain X-ray emission found toward the galactic poles.

IN SOUTH AMERICA, archeologists are searching for primitive records in cave paintings or rock carvings of a celestial explosion that may have appeared a hundred times brighter than the Crab supernova. For one of the largest objects in the Milky Way is the Gum Nebula, which looms across the southern sky. It is a huge cloud of hot gas, probably born of a supernova explosion, so thin that it can be detected only with the most sensitive photographic techniques. Australian astronomer Colin S. Gum, who discovered the cloud in 1952, later mapped it over an expanse 60° by 30° and thought it might reach even as far as the solar system. Gum's studies were tragically ended by his death in a skiing accident in 1960.

The present picture of the Gum Nebula gives it a radius of 1,200 light-years, making it 160 times as broad as the Great Nebula in Orion and reaching to within about 300 light-years of the sun. A great outburst of ultraviolet radiation from the ancient supernova is believed to have flashed through the interstellar gas to produce the "fossilized" nebula—a vast cloud of fluorescent gas that is still glowing as a result of the explosion. Eventually the electrons and protons in the central portions of the nebula will recombine to normal hydrogen, leaving only a faintly

radiating hot outer shell. With an estimated age of 10,000 to 30,000 years, Gum's cloud may be the result of the nearest supernova to have exploded in "recent" times.

Could the occurrence of a supernova have profound effects on life on earth? This question was discussed many years ago by Josef Shklovsky and his colleague, V. I. Krassovsky, who concluded that such events may well have happened many times.

The two suggested the following hypothetical sequence:

A supernova 30 light-years away bursts forth, a thousand times brighter than the moon. Ultraviolet radiation ionizes the earth's high atmosphere to ten times its normal activity, resulting in far more brilliant airglow displays, but no radiation penetrates to the earth's surface. After several years the supernova no longer is visible to the unaided eye, but its nebula continues expanding.

Ten thousand years later the shell of gas reaches the earth and, for the *next* 10,000 years, the earth is enveloped in a web of nebular filaments. Fantastically beautiful luminous streaks cover the sky like brilliant northern lights. At first, swarms of cosmic rays bombard the earth with a hundred times the normal intensity, and for hundreds of years—perhaps as long as 30,000 years—the cosmic-ray exposure continues to exceed by several times the present level.

Shklovsky and Krassovsky speculate that the disappearance of the dinosaurs may have been caused by an overdose of cosmic rays from such a supernova. On the other hand, increased radiation may not always be harmful—indeed, it may have stimulated the evolution of living cells from simple organic compounds billions of years ago.

Expanding gas creates a halo around a hot blue star, forming a planetary nebula in the constellation Aquarius. Late in its life, the central star ejects thin shells of gas, but details of the process still puzzle astronomers.

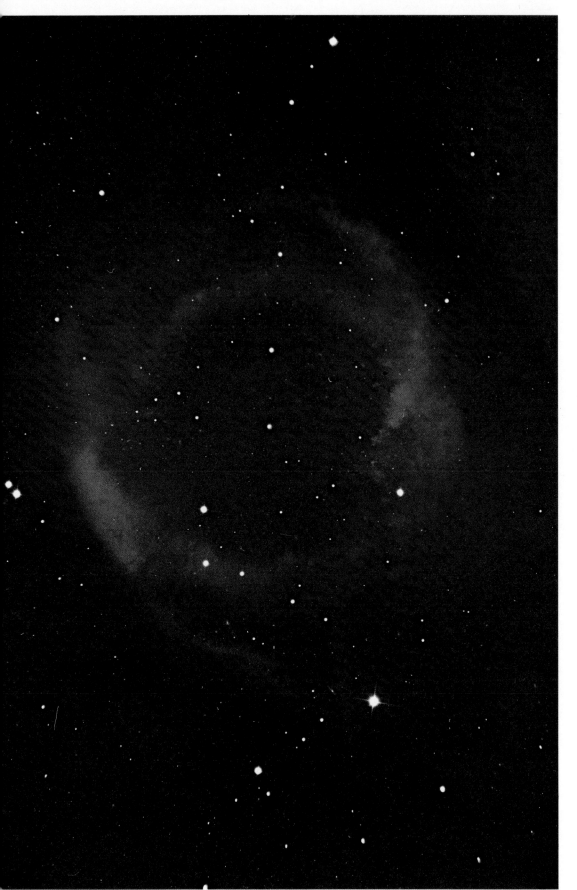

Stellar
Beacons
and
Black Holes

ONE OF THE FASCINATIONS of scientific study is the challenge of the unexpected. The discovery in 1967 of pulsars—neutron stars that are rapidly pulsing sources of radiation—ranks as one of the most surprising events in the history of astronomy.

Anthony Hewish and his colleagues at the University of Cambridge in England were conducting research on the "twinkling" of mysterious, very distant objects called quasi-stellar radio sources, or quasars for short. When photographed through a large telescope, these objects look like faint blue stars, yet they seem to be pouring forth more energy than a hundred large galaxies from the outer reaches of the universe. The Hewish project required a very sensitive radio telescope that could detect changes in the strength of the quasars' radio signals in a fraction of a second.

Such telescopes must be much larger than optical instruments because radio wavelengths are ten thousand to ten million times as long as visible-light wavelengths. Interestingly enough, the radio telescope designed by Hewish was not a complex instrument but consisted mainly of wires strung across a big field.

A young Irish graduate student, Jocelyn

Overleaf: World's largest radio telescope turns a giant ear on space at the Arecibo Observatory in Puerto Rico. Cables hold its 600-ton antenna high above workers inspecting the thousand-foot-wide aluminum reflector.

Bell, was assigned responsibility for construction of the antenna "farm," which covered 4½ acres. The farm was planted with an array of 2,040 small antennas connected by cables; and after two years largely spent swinging a 20-pound sledge hammer, the slight young woman was more than ready to begin operating the new telescope. It had been designed to detect the twinkling, or scintillation, of distant radio galaxies and quasars. As it turned out, its qualities were ideally suited to the discovery of pulsars.

Results of radio reception from one complete survey of the accessible sky were recorded on some 400 feet of chart paper each week. It was Miss Bell's job to scrutinize the charts for signals that were true twinklers, discarding the results of man-made interference such as that radiated by automobile ignitions and other electrical devices. We are aware of similar effects in our homes, such as the distortion of the television picture or the buzzing in a radio speaker that sometimes occurs when another electrical appliance is turned on.

By repeating the sky survey each week, it was possible to distinguish cosmic radio sources from the variable man-made interference. Miss Bell started the survey in July 1967. In October she detected "a bit of scruff" on about a half-inch section of the 400-foot chart—and remembered noticing such an entry before. "When it clicked that I had seen it before," she said later, "I did a double click; I remembered I had seen it from the same part of the sky before. This bit of scruff was something I didn't completely understand [but] my brain just hung on to it...."

Professor Hewish suggested that a search be made with a faster recorder. For several weeks she could find no repetition of the "bit of scruff." Then, abruptly, the signals reappeared as a series of precise radio pulsations almost 1⅓ seconds apart. Miss Bell telephoned Hewish, whose reaction was, "Oh, that settles it; they must be man-made." Such rapid and regular signals had never before been detected from stars.

But they proved to be celestial. Miss Bell called her strange discovery a Belisha beacon—the name for the flashing orange globe that warns British motorists of a pedestrian crossing. The precision of the timing was incredible. Over a period of weeks the regularity of the radio pulsations was more accurate than a timepiece that loses only a fraction of a second in three years. Whatever else this object might be, it was a remarkable clock.

The sharpness of the signals implied that they originated from a source the size of a small planet. The "beeps" seemed so unnatural that the half-serious suggestion was made that they might be messages from a distant civilization, and the research group labeled the source LGM, for "Little Green Men." Hewish, trying to determine whether the signals came from an alien communications system, later said, "I felt compelled to maintain a curtain of silence until this result was known with some certainty. Without doubt, those weeks in December 1967 were the most exciting in my life."

Soon such speculation faded, however, for Miss Bell found similar pulsations right before Christmas. Hewish confirmed the new source in January, and that, said Miss Bell, "removed the worry about little green men, since there wouldn't be two lots signaling us at different frequencies."

The team found and confirmed two more sources later that month, and in February 1968 the discovery of pulsars—objects scientists had not even dreamed of—was announced to the public.

Around the world, astronomers turned toward these first four pulsars, and astronomy students began stringing wires across meadows and fields to build their own radio telescopes.

The development and refinement of radio telescopes since World War II has had a profound influence on the study of the universe—as profound an influence, in fact, as did Galileo's telescope in its time. Before

Galileo, scientific theories about the heavens were based on evidence gathered by the unaided eye. His telescope—and the vastly improved models of today—have changed mankind's conception of the universe.

But optical telescopes are limited to observations of visible objects. Radio telescopes, free of such restriction, have opened new windows on the sky. Tuning in on noises from the stars, they achieve remarkable sensitivity with relatively little interference from the atmosphere.

Just as improvements in optical telescopes showed that some "stars" were actually entire galaxies, improvements in radio telescopes are helping us explain remarkable celestial objects whose nature was unsuspected even 20 years ago.

Moreover, these listening posts are fulfilling a prophecy made more than a century and a half ago by Herschel. In January of 1820, he and seven other men met in a London tavern to found the Royal Astronomical Society. For his first draft defining the group's objectives, he wrote: "... it is possible that some bodies, of a nature altogether new, and whose discovery may tend in future to disclose the most important secrets... of the universe, may be concealed under the appearance of very minute single stars no way distinguishable from others of less interesting character...." For some reason Herschel decided to delete the passage from his final draft. Perhaps it was just as well, for 150 years would pass before his prediction would be proved correct.

THE SHARPNESS of a radiated signal is related in a simple way to the size of the region that produces it: Because of the time it takes light or radio waves to travel, a large spherical source sends us a fuzzy signal. If, for example, the sun were suddenly to flash off, how would it look to us on earth? We would first see the center of the disk—the part nearest to us—go black. Since the edge of the disk is a little more than 2¼ light-seconds (435,000 miles) farther from earth

HERMAN ECKELMANN, CORNELL UNIVERSITY

Remote from earth-based electrical interference, Arecibo listens to radio noise from the skies. The radio telescope "can carry us to other worlds and to parts of the universe which we could never imagine," says Frank D. Drake (above), director of the observatory. Focusing on our own solar system, the telescope has revised old notions, revealing that Mercury rotates on its axis every 59 earth days instead of 88 as once believed, and that Venus spins clockwise instead of counterclockwise as the other planets do. Technicians (below) wear rubber foot pads to protect the newly installed aluminum panels of the reflecting dish. Replacing steel mesh, the panels have greatly increased the sensitivity and accuracy of the telescope.

NATIONAL GEOGRAPHIC PHOTOGRAPHER ROBERT W. MADDEN (ABOVE AND OPPOSITE)

than the center, we would see the dark center quickly spread until the bright ring at the edge of the disk vanished—about 2¼ seconds later.

So sharp were the signals observed by the Cambridge astronomers that they estimated the diameter of the radiating body in each case to be no more than a few thousand miles. Since the pulsations showed none of the variations that planetary motion would cause, only white dwarf stars and neutron stars could fit the description. And no neutron star—although the concept of such a body had been carefully worked out—had ever been identified.

When the Cambridge results were published, theorists promptly responded, suggesting a variety of possible sources. All of these fit into essentially two categories: vibrating stars and spinning stars. The vibrating or oscillating stars known as Cepheid variables, for example, go from dim to bright and back to dim in days. A star the size of the sun could not brighten and dim in less than an hour or so. But calculations for white dwarfs and neutron stars showed that their vibrations might be fast enough to cause the strange pulses found by the Cambridge astronomers. It was difficult to understand how oscillation of the surface of a star could generate radio waves, but one explanation was that the surging motion of the surface would cause shock waves, and these in turn could generate radio waves as they disturbed the star's outer atmosphere.

Thomas Gold of Cornell University, seeking to explain the pulses as a result of star rotation, used a lighthouse analogy. Perhaps the radiation emerged from some active region on the spinning star, forming a beam. If the star and the earth were in proper alignment, the beam would sweep across the earth, flashing or pulsing with each stellar rotation. For a star as large as a white dwarf, this presents difficulties; a white dwarf spinning fast enough to send a beam to the earth every second would tear itself apart. But a neutron star as small as ten

miles in diameter could spin as fast as 600 revolutions per second before centrifugal force would destroy it.

Just as a figure skater starts a spin with arms outstretched and whirls faster as the arms are brought in close to the body, so stars spin faster as they collapse. Both the collapse and the increase in speed are dramatic. A star that before collapse rotated once in several days, after its collapse suddenly whirls around in a matter of seconds.

Similarly, the magnetic field of a collapsing star would be sucked into a very high concentration. The initial magnetic field would increase ten billion times at the surface of the neutron star. The spinning neutron star would have incredible magnetic strength. Looping lines of force would stretch from north pole to south pole to form a strong magnetic cage, or magnetosphere.

In Gold's view the pulsar was a sort of cosmic slingshot. Any ionized gas, or plasma, produced near the surface of the star would be propelled along the bars of the magnetic cage. As they traveled away from the surface, the electrons trapped by the star's powerful magnetic force would move faster and faster. As they approached the speed of light, the electrons would generate a beam of synchrotron radiation—radio waves, visible light, perhaps even X rays.

SINCE THE CAMBRIDGE discovery, about 150 pulsars have been found. The fastest of them was detected in 1968 in the Crab Nebula. It pulsed its radio signals at the astonishing rate of 30 per second. Certainly it could not be larger than a neutron star. Astronomers rushed to explore every other range of the electromagnetic spectrum for further evidence.

To optical observers, this was an opportunity to train their large telescopes directly on the suspected source, one of two faint stars close to the center of the nebula. Almost immediately they detected visible-light pulsations. They had found a star, brighter than the sun, flashing on and off 30 times a second!

My colleagues and I placed instruments aboard an Aerobee rocket in the early spring of 1969 to search for X rays from the Crab. We found, to our great surprise, that it radiated ten thousand times as much pulsed power in X rays as it did in radio signals. Altogether, the pulsations of the Crab have a power ten thousand times that of the total heat and light pouring from the sun.

Our discovery of X-ray pulsations from the neutron star in the Crab coincided with a symposium on pulsars at the National Academy of Sciences. I hurried to include a description of our results in a session led by Tommy Gold.

Soon after our conference, astronomers at the radio observatory at Arecibo, Puerto Rico, issued a report concerning the Crab's pulsar clock. Although its timing had at first seemed perfect, very exact measurement had shown that it loses 15 millionths of a second in a year.

When this news reached Cornell, Gold made one of his typical quick calculations of the energy that would be dissipated in the slowing down of the neutron star. He then called me at the Naval Research Laboratory to ask a crucial question: What was the total power radiated in X rays by the star? At my answer he exclaimed with excitement that the fit was perfect! The energy released by the slowdown in the spin of the pulsar was an incredible ten thousand trillion trillion kilowatts. The transformation of that tremendous energy into radiation could account for all of the star's radio, visible, and X-ray power!

Here was the answer to the riddle of the Crab—a story of cosmic death and transfiguration. The original star, which presumably burned a normal lifetime of billions of years, had died in a violent collapse and explosion, then had been reborn as a pulsar. Nearly all of the original star's vast gravitational energy was retained by the pulsar.

This will enable it to radiate pulses of energy at very high power for thousands of years to come.

A NEUTRON STAR challenges all our understanding of matter at ultrahigh densities. David Pines and Mal Ruderman have been intrigued with the structure of neutron stars and have suggested interesting comparisons with the earth. A neutron star, of course, is fantastically more compact. If the earth were similarly compressed, it would measure only 600 feet across.

Both the earth and the neutron star have solid crusts, liquid interiors, and, possibly, solid central cores. In the neutron star, however, the surface temperature may be more than a million degrees, and the crust trillions of times as strong as steel, with a melting point of trillions of degrees. Inside the crust the temperature is several billion degrees, and the density is a billion tons per cubic inch.

Nearly all the pulsars now known are very gradually slowing down. But the two fastest pulsars—in the Crab Nebula and in the constellation Vela—frequently stutter and, once or twice a year, jump abruptly in frequency. Pines and his colleagues attribute these events to starquakes. As the spinning star slows down, its crust is subjected to enormous stress, and eventually a crack opens up like that sometimes caused by an earthquake. This abruptly relieves the stress; the star shrinks slightly and speeds up for a time. Then it settles back to its normal rate of slowdown until another starquake occurs.

Timing a pulsing star from a distance of 5,000 light-years, astronomers can measure a speedup of even one part per billion, and thus easily detect a starquake that results in a shrinkage of only a few thousandths of an inch. Freeman Dyson has commented that the seismic study of neutron stars should present as "rich and bewildering a variety of phenomena as exists in the mantle of the earth."

OF ALL THE RADIO PULSARS that have been discovered, not a single one other than the Crab clearly pulses at the shorter wavelengths of visible light and X rays. But a large class of X-ray sources is made up of pairs of stars that orbit each other. In such double or binary stars, one partner is most likely a neutron star and the other a normal star. A wind, or stream of gas, flows from the normal star toward the compact neutron star. Thus matter plunges from one to the other with such energy that the temperature rises to tens of millions of degrees, and the heated gas emits thermal X rays.

More than half the stars in our Milky Way galaxy are members of multiple systems in which two or more stars orbit one another. If the position of a binary is right in relation to the earth, we see the two stars eclipse each other at regular intervals.

Some of the earliest observations of eclipsing binaries were made by a young deaf mute, John Goodricke, son of the Baron of York. While still a teen-ager he won recognition as a skillful and accurate atronomical observer, and in 1782 measured the fluctuations of the variable star Algol. From careful study of its light variations, he conjectured that it could be an eclipsing binary. William Herschel at first disagreed, but later had to admit that Goodricke was right. Today, Algol is studied for its violent radio flares, which often last for weeks.

Although Goodricke died when he was only 21, he had several major discoveries to his credit, among them the fact that the huge star Beta Lyrae was an eclipsing binary varying from bright through dark to bright again in a period of 12 days, 21 hours, 48 minutes. This binary system has turned out to be among the most fascinating objects in the sky, and numerous astronomers, including the late Otto Struve, have dedicated much of their effort to studying it. A large amount of matter is constantly being transferred from one of the member stars to the other, and much also is escaping the binary system; as a result, its orbital period is slowly changing. One of the pair is visible, but its companion, believed to be the more massive of the two, has not been detected optically, and this has puzzled astronomers for decades.

Recently an international team of astronomers, conducting observations with the NASA-Princeton University Orbiting Astronomical Observatory *Copernicus*, found that Beta Lyrae is surrounded by extensive hot, highly ionized gas clouds that dominate its ultraviolet spectrum. All evidence indicated an enormous flow of matter between the two member stars.

Binary systems whose partners are interacting in their evolutionary processes are called close binaries. Their fluctuation periods are typically on the order of days, but range from a few hours in the case of two dwarf stars to tens of years for a system involving two supergiants.

In a well-separated system, each star remains nearly spherical except for some flattening at the poles from rotation. In a closer pair, they tend toward egg shapes, with the more sharply curved ends facing each other. So-called "contact systems" sometimes develop when the more massive member starts expanding outward in its evolutionary process, because of faster burning of its nuclear fuel.

As the hydrogen in the stellar core is converted to helium, nuclear fusion takes place farther and farther out from the core—the process described in tracing the life cycle of our sun. The star begins to expand to become a giant or, perhaps, a supergiant.

What happens if the more massive partner becomes too large in relation to the separation of the two stars? Applying celestial mechanics, astronomers learned long ago that, in a given binary system, there exists a surface called the critical Roche lobe beyond which neither star can grow without spilling its gas throughout—and probably out of—the binary system. If the egg-shaped star reaches the gravitational balance point between the two stars, known as the inner Lagrangian point, the star has approximately filled the critical Roche lobe.

Just how the gaseous mass will start pouring out of the evolving star is still not well understood; this is partly because the forces operating here are not limited to the gravitational attraction between the two stars, but also include gas pressure, stellar wind, radiation pressure, and perhaps other factors. Part of the material may leak out of the binary system through the second and third Lagrangian points, located outside the stars on a line joining their centers. In the course of the evolutionary processes of both stars, they may fill up their respective critical Roche lobes by transfer of matter from one star to the other, and during the lifetime of the system this transfer may take place back and forth several times.

AMONG THE BEST-STUDIED X-ray eclipsing binaries is that made up of HZ Hercules and Hercules X-1. The visible member, HZ Hercules, has long been known to astronomers as a variable star. The X-ray partner, Hercules X-1, was discovered in 1971 by Riccardo Giacconi and his colleagues from observations made with the astronomy satellite *Uhuru* — the first all-X-ray satellite, whose name is the Swahili word for freedom. It was launched on December 12, 1970, from Italy's San Marco Test Range in the Indian Ocean, off the coast of Kenya and near the town of Malindi. Most of the community watched the lift-off from bleachers set up at the base camp at the ocean's edge. The day, the event, and the satellite's name celebrated the seventh anniversary of Kenyan independence — and also the new freedom of X-ray astronomers from the interference of the lower atmosphere.

In the Hercules binary, optical telescopes reveal HZ Hercules to be a blue star almost twice as massive as the sun. The system's X rays come from the invisible companion, Hercules X-1. The unseen star behaves like a pulsar, but it sends out no radio waves. The rapid X-ray pulsation and the computed mass suggest that Hercules X-1 is a neutron star.

In this system, it appears that the normal blue star is spilling gas through the inner Lagrangian point onto the neutron star. The heat of impact raises the temperature of the neutron star to tens of millions of degrees, and thermal X rays are emitted from the polar regions; as the star spins, its poles come into view of X-ray detectors aboard rockets or satellites once every 1.24 seconds.

ARE NEUTRON STARS the ultimate limit of crushed matter, or can collapse go even further? It is this question that has led to the intriguing concept of the "black hole."

Suppose more weight is piled onto a neutron star than it can support. In such an event there would be no way to hold off total, catastrophic collapse. When a star three times as massive as the sun has shrunk to about 11 miles across, the force of its gravity becomes so powerful that anything escaping its surface would have to exceed the speed of light, and nothing — *not even light* — can escape. All external evidence of the star disappears, leaving only its disembodied gravity to mark the black hole in space. I am reminded of the Cheshire cat in *Alice In Wonderland*, which faded away entirely except for its grin.

Inside the black hole, matter crushes down to greater and greater density. In its rush to oblivion, the star theoretically shrinks past pinpoint size to submicroscopic, and finally on down to zero volume. Of this final condition, physicists say simply, "We don't know what happens ultimately." It is difficult, of course, to accept such an astounding theory, yet we know of no physical force to prevent it. Perhaps some new law of physics, some new property of nature will eventually reveal itself to write the end of the story of total collapse and open up visions of an entirely new, presently unimaginable, universe.

LICK OBSERVATORY (BELOW)

PAINTING BY DAVIS MELTZER

JONATHAN S. BLAIR (OPPOSITE AND BELOW)

MULLARD RADIO ASTRONOMY OBSERVATORY (BELOW)

Cosmic Lighthouse

Array of antennas stretches beyond Anthony Hewish at the Mullard Radio Astronomy Observatory in Cambridge, England. In 1967 the receivers picked up mysterious, precisely timed radio pulsations. Jocelyn Bell Burnell (upper right), then a graduate student of Professor Hewish, first noticed the radio signals — represented by the peaks in the recording above. Dr. Hewish reasoned that the strange pulses could be man-made interference, an unknown natural occurrence, or even a signal from beings of another world. Half-seriously he called the source LGM, for "Little Green Men." Ultimately, astronomers determined that the radio pulses origi-

nate from a natural source, a rapidly spinning neutron star no more than a dozen miles in diameter. Called a pulsar, such a tiny star forms after the collapse of a vastly larger star. The pulsar retains most of the material of the original star, and a handful of it would weigh billions of tons.

A pulsar's extremely rapid spin and intense magnetic field (curved blue lines) combine to beam narrow bands of energy (opposite) like a rotating beacon in a lighthouse. The arrow indicates the direction of spin, red lines the axis. A visible pulsar in the Crab Nebula blinks on (far left, upper) and off (far left, lower) 30 times a second.

Consideration of the black-hole enigma brings to mind various other unfamiliar concepts, such as curved, or warped, space and trapped light.

We are accustomed to thinking of space in dimensions that permit us to determine and describe position and direction in familiar, predictable ways. But in the neighborhood of a collapsing star, space becomes curved. Think for a moment of the flat surface of a stretched rubber sheet. If a billiard ball is placed at the center, the sheet will sag in a deep dimple. On a fixed surface of this shape, a ball rolled in the general direction of the hole — but not straight at it — will dip toward the hole, then roll away at an angle to its original direction. If it starts out only slightly off the mark, it will be drawn to the depression and will spiral in.

Space surrounding a star is curved in a similar way, and a ray of light is affected like the path of the rolling ball. In such conditions light can be bent or trapped.

Near the earth the curvature of space is so slight that we cannot detect the smallest deviation in the paths of light rays. Even near the massive sun the curvature is very slight, but it is detectable. In 1916 Einstein predicted that a ray of starlight passing close to the edge of the sun would bend slightly. Because the sun is so dazzling, this observation could be made only at the time of a total solar eclipse, when stars close to the sun could be seen against a dark sky.

Einstein made his prediction in the midst of World War I, so it was not until the solar eclipse of 1919 that scientists could test his theory. During the final months of the war, astronomers pushed preparations for expeditions to Brazil and West Africa. They described their mission as an experiment to "weigh light," because — Einstein's theory proposed — a ray of light passing the sun would bend toward the sun under its own weight just as the path of a bullet bends toward the ground.

Eclipse conditions in Brazil were ideal. The results of the test agreed with Einstein's predictions, and a new era in scientific thought began. Over the years, the eclipse test has been repeated a number of times with similar results, and recent measurements of radio sources have confirmed the theory to an accuracy within one percent.

Gravity is normally the weakest natural force in the universe, a trillion trillion trillion times weaker than the force between subatomic particles in ordinary matter, such as stone or wood. Ordinary matter contains an equal mixture of protons and electrons; the protons have a positive charge and the electrons a negative charge. The attractions and repulsions of these particles balance so perfectly that we do not sense the tremendous forces at work. Professor Richard P. Feynman impresses his students at the California Institute of Technology with this illustration: "If you were standing at arm's length from someone and each of you had one percent more electrons than protons, the repelling force would be incredible. How great? Enough to lift the Empire State Building? No, more than that! To lift Mount Everest? More than that! The repulsion would be enough to lift a 'weight' equal to that of the entire earth!"

But when a large mass collapses, its gravity overwhelms all other forces. If the earth were to collapse from a diameter of nearly 8,000 miles down to 2,000 miles, its increased gravity would cause a 175-pound man to weigh almost 1½ tons. If it were squeezed to a diameter of only two miles, the man's weight would increase to about 1½ *million* tons. At a diameter of about 2/3 of an inch, the earth would become a black hole — if the theory can be trusted that far.

Only months after Einstein published his theory in 1916, German astronomer Karl Schwarzschild solved the mathematical problem of the curving of space by considering what happens in the gravitational field of a star collapsing to what we now call a black hole. Space close to the star becomes so warped that it curves in on itself completely. Gravity still exerts its force outside the

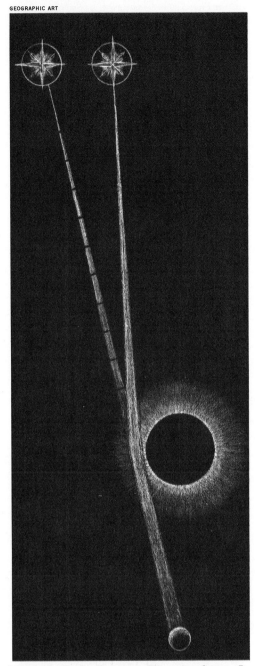

collapsing star, but no light can emerge from inside. The boundary of the entrapping region is called the "event horizon."

Since nothing can escape from a black hole—not even light or X rays—there is no way an observer investigating the immediate vicinity could report back to us. But if he could somehow approach the black hole, he would first encounter a "photon sphere." Near the photon sphere, light rays passing the black hole are bent from their straight paths. Those rays that skirt the hole at a distance are only slightly deflected, but close to the black hole, rays are more and more severely curved. Finally, within the photon sphere, a ray of light enters into a circular orbit: Looking straight ahead in the photon circle, the observer theoretically would see the back of his head.

Once inside the photon sphere, the observer encounters the phenomenon of the "exit cone." Suppose he is stationed on the surface of the collapsing star, equipped with a searchlight. He points his light beam directly overhead, and the rays travel off into space. As he tilts the searchlight slightly away from the vertical, the rays start to bend, though they still escape. But then, at larger angles, the rays are trapped by the intense gravity, curving all the way back to the star's surface. The exit cone narrows rapidly and finally closes up completely at the event horizon. Inside, collapse continues inexorably to a conclusion that has meaning only in theoretical terms: zero volume and infinite pressure and density.

To understand the environment of a black hole, there is one other major aspect of Einstein's theory that should be considered: the effect of an extremely strong gravitational field on time, the so-called fourth dimension. Just as gravity can affect matter and even light, it can affect time by slowing down all the indicators of the passage of time—not only the frequency of vibrations of an atomic clock, for example, but also the life processes of a hypothetical observer approaching the black (Continued on page 128)

"Optical illusion" proves Einstein theory. In 1916 the famed scientist proposed that light rays bend when they encounter a powerful gravitational pull such as that of the sun. An experiment during the solar eclipse of 1919 dramatically supported his point: Scientists measured the position of a star as the sun came close to the "line" between the star and earth. Solar gravity deflected light from the star (solid line) and the apparent position of the star changed. This gave the illusion that the star lay in a direct line from earth (dashed line).

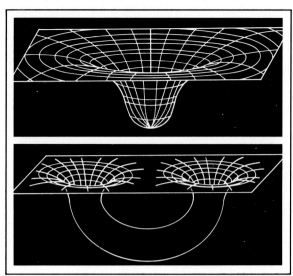

FROM "RELATIVITY AND COSMOLOGY" BY WILLIAM J. KAUFMANN, III; HARPER AND ROW, 1973

Warped Space, Trapped Light

Seeking the cause of X rays from a mysterious and invisible source, physicists John A. Wheeler of Princeton University (upper left, at left) and Peter Goldreich of the California Institute of Technology discuss the magnetic fields between two stars orbiting each other. They searched vainly for a source other than a black hole — a star that has collapsed and crushed itself into invisibility by its own intense gravity. Particles pulled toward a black hole compress and heat to tens of millions of degrees, giving off X rays before they vanish into the hole. Studying the results of Uhuru, the first orbiting X-ray satellite, Riccardo Giacconi of the Smithsonian Astrophysical Observatory (far left) identified a possible black hole in 1973. It lies in a double-star system called Cygnus X-1.

Six decades ago Einstein predicted that space warps around collapsed stars, as in the sinkhole (upper diagram), and wraps completely around a black hole to isolate it from the rest of the universe. Stranger still, the theory of the wormhole (lower diagram) proposes that everything drawn into a black hole at one end will spurt out another hole, a white one, somewhere else in our universe — or even in another universe. Supermassive black holes may form at the centers of galaxies and quasars. Thus speculates Kip Thorne of the California Institute of Technology (left), displaying a model of space as it might curve around a black hole.

Martin Schwarzschild of Princeton (center) has researched the life cycles of stars, laying the foundations for theories about the eventual collapse of stars.

DANA DOWNIE (BELOW)

The Black Hole

Gravitational whirlpool, possibly caused by a black hole, swallows up gas from its companion star in an illustration of the double-star system known as Cygnus X-1 (opposite). The gases whip around the hole in a tight circular orbit to form a thin accretion disk (diagram, below). Extremely hot particles give off X rays while being dragged toward the hole. Some scientists believe that such holes complete the evolution of stars that once had many times the mass of the sun.

The gravity of blue supergiant stars carries them through several stages of nuclear burning of successively heavier elements. Finally, they collapse to neutron stars about ten miles in diameter, or possibly to black holes that disappear from the observable universe.

If the earth, while retaining its present mass, should shrink to about half the size of a golf ball, it would become a black hole.

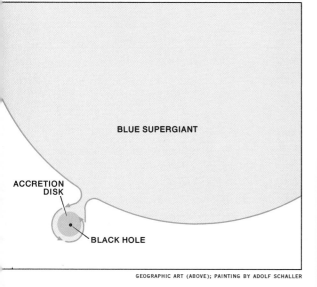

BLUE SUPERGIANT

ACCRETION
DISK

BLACK HOLE

GEOGRAPHIC ART (ABOVE); PAINTING BY ADOLF SCHALLER

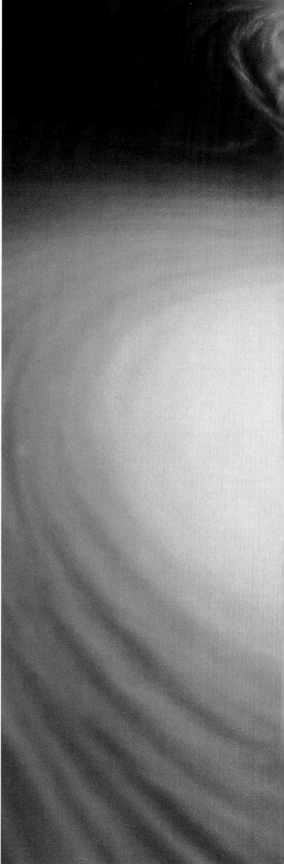

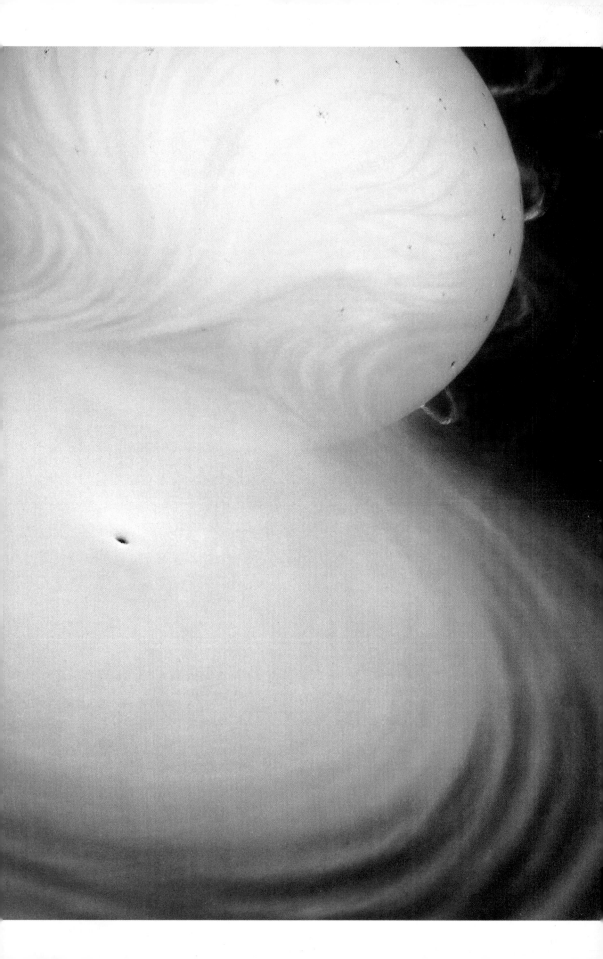

hole. The observer senses no slowdown, since his watch, his heartbeat, and his other functions slow down in synchronization with one another. But imagine that we can set our clock with the watch of an observer on the surface of a collapsing star with a mass three times that of the sun. He would read an elapsed time on his watch of about a second from beginning of collapse until he plunges into the entrapping region. But if we looked at the watch with a telescope, our observations might go as follows: When the diameter of the shrinking star is a thousand miles, the timepiece on the star would be losing about seven minutes per day; at a diameter of 30 miles, it would lag about five hours per day. At the event horizon, with the star's diameter diminished to about 11 miles, the watch on the star would stand still.

As time comes to a halt, the collapse of the star appears to us to stop; and it grows rapidly dimmer and redder because of two combined effects. First, as the exit cone narrows down to a smaller and smaller opening, more and more of the light trying to escape is bent back — into the black hole. Second, as gravity stretches the wavelengths of the radiation coming from the star, its color passes through the red range — the "gravitational red shift" — to invisible infrared. Abruptly the star disappears from our view.

ALL OUR DISCUSSIONS of black holes, of course, have to remain theoretical, and the observers we describe remain hypothetical adventurers. In the words of Kip Thorne of the California Institute of Technology, "You will never be able to see inside a black hole, and you can never know what has happened inside, since no energy in any form ever comes out to carry the information. Roger Penrose at Oxford calls it 'cosmic censorship.'

"Yet it is easier to analyze the surface of a black hole than that of a star. We believe a black hole is an extremely smooth structure; it can never have ripples or mountains. Anything it traps can never escape. At Cambridge, Stephen Hawking has proved that the black hole can neither split nor decrease in size; it can only grow, and nothing can prevent it from growing.

"Ultimately, if the universe itself does not collapse and die first, the black holes will eat up all the matter in our galaxy."

How can astronomers find a black hole in the sky? Even one just a few light-years from earth would give no visible indication of its existence, except under very special and rare circumstances.

Trying to arrive at a method of searching out black holes, two Russian astrophysicists, Y. B. Zel'dovich and O. K. Guseynov, theorized about what would occur if a black hole and a visible star were revolving around each other. As the visible star orbits the black hole, its spectral lines should redshift and blue-shift alternately, as it moves away from and then toward us in each revolution. The plan of search, therefore, was to find binary systems in which the spectral lines of only one star could be detected. Thousands of such systems exist, but the search was narrowed to those in which the mass of the invisible partner was at least three times greater than that of the sun. The larger the mass of the black hole, the greater would be the pull on the visible star, thus increasing the extent of the shifts.

Unfortunately, even though the candidates numbered only half a dozen, it was possible in each case to come up with explanations other than a black hole to account for the invisibility of one of the stars.

Only one promising approach was left. If the black hole sucked gas from its companion star, the rush of matter toward the hole could release X rays. If in an X-ray binary the mass of the invisible member was more than three times that of the sun, then — theoretically — it would be a black hole.

One excellent candidate for such a condition is Cygnus X-1. The evidence concerning its visible companion, a blue supergiant, and the orbital characteristics indicate that the X-ray star has a mass more than six times that of the sun.

The concept that emerges from theory is of a stream of gas that pours from the visible star through the Lagrangian point and strings into orbits around the black hole. Centrifugal forces flatten the flow of gas into the shape of a thin disk—like the rings of Saturn on a much larger scale. Friction within the disk forces the gas to spiral toward the black hole. Within a matter of weeks or months, gas works its way several million miles from the outer edge of the disk to the inner edge, where the overwhelming pull of the black hole finally captures it. By the time the gas has spiraled to the edge of the hole, friction has raised its temperature to tens of millions of degrees. The hot, swirling gas pours out X rays just before it plunges to oblivion inside the black hole.

The innermost portions of the disk must become very turbulent, and the X rays accordingly should fluctuate in intensity. Here, perhaps, is a key to identification of a black hole with the X-ray emission of Cygnus X-1. For it was our earliest example of a variable X-ray star. After my colleagues and I had located the two brightest X-ray sources, Scorpius X-1 and the Crab Nebula, in 1963, we hurried to scan the broad sky with much greater sensitivity. Larger X-ray detectors were mounted to look out the side of an Aerobee rocket. By slightly canting the tail fins, we could induce a slow spin as the rocket rose through the lower atmosphere. Once beyond the effects of air pressure, the roll of the rocket expanded into a wobble that permitted the detectors to scan from zenith to horizon in all directions.

In 1964 we discovered several new X-ray sources, of which Cygnus X-1 and Cygnus X-2 were the brightest. Scorpius X-1 and the Crab were unchanged from 1963. Having recovered the payload by parachute, we repaired it and flew it again in 1965. Now we found a fourfold decrease in the brightness of Cygnus X-1, while Cygnus X-2 remained unchanged. Such a large variation

in so strong a source was difficult to believe, but the evidence was indisputable. As more and more sky surveys were made by rocket and balloon and finally by the satellite *Uhuru*, it became clear that X-ray stars were typically highly variable; the constant star was indeed a rarity.

The theory of the turbulence of the disk of Cygnus X-1 predicts very rapid fluctuations. And recent X-ray observations by Elihu Boldt and his colleagues at the Goddard Space Flight Center in Greenbelt, Maryland, do indeed show as much as fivefold fluctuations in intensity—very rapid staccato bursts, each occurring in less than a thousandth of a second.

In the mid-1970's, a new class of X-ray sources has been rather convincingly identified with globular star clusters in the Milky Way. One possible explanation is that very large black holes, of about a thousand times the mass of the sun, exist at the centers of clusters. Perhaps the most massive stars in the clusters evolved to black holes eons ago. After billions of years, some theorists reason, the individual black holes would sink to the center of gravity of a cluster and finally coalesce into one vast black hole. X rays are emitted when gas drawn toward the black hole gets very hot. Such gas may be supplied by the accumulated debris of stars torn apart by tidal forces when they happened to wander too close to the black hole.

Do black holes really exist? Astronomers are still very uncertain. Says Kip Thorne, one of the leading theorists involved, "I'm about 80 percent sure that Cygnus X-1 contains a black hole. Future confrontations between observations and theory may strengthen that conviction—or may destroy it."

The answer may not be far off. The next major spacecraft being prepared for X-ray astronomy — HEAO (High Energy Astronomy Observatory)—will be specifically instrumented to search for such black-hole evidence as extremely fast X-ray fluctuations.

SINCE THE LATE 1950's, Joseph Weber at the University of Maryland has been searching for evidence of gravitational waves predicted by Einstein. Such waves might signal the collapse of stars to black holes, or possibly the swallowing up of stars by a supermassive black hole in the center of the galaxy.

A fundamental precept of Einstein's General Theory of Relativity is that nothing can travel faster than light, whereas Newton's gravitational theory had presumed no limiting speed: The effect of forces acting between bodies was instantaneous, regardless of distance. For example, if the sun were somehow suddenly abolished, the earth would immediately take off in a straight line through space.

But according to Einstein's theory, such a disappearing mass would change the curvature of the space around it, and transmit gravitational waves. The waves would travel at the speed of light and should carry an amount of energy equal to that of the disappearing mass.

Weber's detector is a large, solid cylinder of aluminum suspended on fine wires in a vacuum. Arrival of gravitational waves should set the cylinder in vibration. The expected effects are extremely slight, and Weber uses two cylinders — one at College Park, Maryland, and the other near Chicago — so he can distinguish simultaneous signals from spurious noise in each detector. He believes he has detected evidence of a fantastic source of gravitational waves at the center of the Milky Way — perhaps a black hole so vast that it gobbles up hundreds of stars per year. But other experimenters have not been able to reproduce Weber's results, and so, for the time being, the study must be interpreted with extreme caution. Nevertheless, his research has encouraged not only other bold experiments but also the development of useful new instruments and tools.

After the Nuclear Test Ban Treaty of 1960, the United States designed a system of satellites that could detect the X-ray flashes of any secret weapons test in space. No such violation has been discovered, but the Vela satellites have collected much data on solar flares and the earth's radiation belts.

In 1967 Ian B. Strong and his colleagues at Los Alamos became aware of a most unusual type of signal in their records from the satellites. Lightning-like bolts of X rays or gamma rays of very high intensity and extremely short duration were detected at the same instant in several satellites widely separated in space. If such signals came from any one instrument, it would be suspected of some malfunction. But their simultaneous appearance in two or more satellites made it clear that these were real events.

About half a dozen such cases have been recorded in each of the last several years. Each burst lasted a few seconds or tens of seconds. The sources are clearly outside the solar system, but may be either in our galaxy or beyond it. If the sources lie within the Milky Way, the energy of each burst is a thousand times greater than the output of the largest solar flares. If the sources are in distant galaxies, the energy released may be equal in power to that released by an entire galaxy in the same time period.

Naturally, these remarkable signals have given rise to a host of speculations. In some ways they resemble the outbursts of giant flare stars and the predicted flashes of exploding stars. Another explanation suggests collisions of comets with neutron stars or black holes. Still another proposes that miniature black holes exist, occasionally become unstable, and erupt in fireballs.

There are major drawbacks to each of these suggestions, and of course it would not be surprising if all turned out to be wrong. Indeed, the truth may lie beyond our present imagination.

Technicians install an X-ray sensor in the Apollo spacecraft for the joint Apollo-Soyuz orbital mission of July 1975. The instrument detected radiation from collapsed stars and the debris of old stellar explosions.

Realms
of the
Galaxies

varying degrees of forking and shredding. The Milky Way, like other spiral-shaped galaxies, is a balanced community of stars at all stages from birth to death. Its bulge is filled with old red stars; its spiral arms are defined by young blue stars, gas, and dust.

Elliptical galaxies, on the other hand, range in shape from nearly spherical to flattened ovals, and are almost devoid of the gas and dust that form new stars; only old stars remain. The irregularly shaped galaxies contain mostly young stars.

A small proportion (about two percent) of the galaxies belong to a special subdivision of spirals, a class discovered by Carl Seyfert in the 1940's. They have very bright nuclei and dim, rather poorly developed spiral arms, so they look deceptively like stars in short photographic exposures. The centers appear to be about ten light-years in diameter, far smaller than the center of our galaxy; yet the total output of energy of one of these Seyfert galaxies may exceed that of the Milky Way a hundred times. The greatest radiation is in the infrared.

The Milky Way is a member of a small "local group" of about 20 galaxies that spans a region three million light-years across. The Andromeda galaxy and the Magellanic Clouds belong to the same cluster. Thousands of such groups can be identified in the National Geographic Society—Palomar Observatory Sky Survey. Sixty million light-years away in the constellation Virgo lies the nearest "rich" cluster, containing hundreds of galaxies. The nearest "great" cluster, with thousands of members, is in Coma Berenices at a distance of 400 million light-years. George O. Abell has catalogued nearly 3,000 great clusters out to distances of about four billion light-years.

Giant galaxies are generally found near the centers of rich clusters. A supergiant elliptical may contain more than ten trillion stars, and measure 300,000 light-years across.

The sheer immensity of such systems suggests eternal qualities of stability and predictability. It *(Continued on page 140)*

BEYOND THE MILKY WAY reaches the seemingly endless realm of the galaxies. Vast though our own galaxy is — 100,000 light-years in diameter, and containing about 200 billion stars — it is just one of perhaps ten billion separate star systems extending as far as we have been able to search with our most sensitive instruments.

The exploration begun by optical means has been greatly extended by radio astronomy, not only in the study of ordinary galaxies but also in pursuit of the quasars — those mysterious objects once called "radio stars" but increasingly thought to be very distant and incredibly powerful galaxies.

Whereas individual stars condense or expand as spheres, galaxies come in a variety of shapes. Edwin Hubble, working with the 100-inch telescope at Mount Wilson — largest in the world from 1918 to 1938 — established the stellar nature of these systems and studied their form and content. Of the 600 galaxies he classified, he found most to be spiral-shaped, many to be elliptical, and a few to be irregular.

The spirals resemble enormous whirlpools, with arms that unravel from a central bulge and coil around it in a thin disk with

Overleaf: Radio telescopes, like this one near Bonn, West Germany, listen for a wide range of celestial signals, from exploding galaxies to quasars — mysterious starlike sources of radio waves, possibly in the farthest reaches of space.
NATIONAL GEOGRAPHIC PHOTOGRAPHER JAMES P. BLAIR

COURTESY OF BELL LABORATORIES

Beginnings of Radio Astronomy

Pioneer of radio astronomy, Karl G. Jansky in 1932 accidentally discovered radio waves coming from space. Assigned by Bell Laboratories to pinpoint the source of static interfering with transoceanic radio-telephone service, Jansky built the first radio telescope (above) — a rotating antenna array mounted on four Model-T Ford wheels to carry it around a circular track. At left, he indicates the position of the first radio noise he detected — a strange hissing sound pouring from the center of our Milky Way. Realizing the importance of his find, Jansky proposed a large dish-shaped antenna with a sharper beam. Although he could convince no one to build the antenna, his work gave birth to the science of radio astronomy.

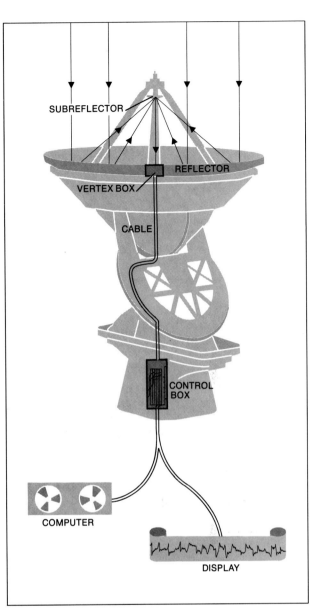

SUBREFLECTOR

REFLECTOR

VERTEX BOX

CABLE

CONTROL BOX

COMPUTER

DISPLAY

Tuning In on the Skies

Far from man-made radio noise, a cluster of telescopes at Green Bank, West Virginia — part of the National Radio Astronomy Observatory network — listens for weak radio signals from the sky. Radio telescopes, though varying greatly in size and structure, all operate by focusing radio waves. The 140-foot-wide telescope at Green Bank, using a curved reflector, gathers radio waves beaming through space and bounces them up to a subreflector (diagram). The subreflector mirrors them into a vertex box, which amplifies and converts the radio waves to electrical signals. Further amplified and converted in the control box, the signals appear as a graph on a display. A computer records and translates the signals into useful information.

Beyond the Milky Way

Galaxies beyond the Milky Way vary in shape from ellipticals (below) to spirals (opposite, top and bottom rows) to irregulars like the ragged, featureless Small Magellanic Cloud (opposite, middle). In 1924 astronomer Edwin P. Hubble, who classified these celestial objects according to structural form, proved that they were external galaxies — large systems of stars and nebulae — comparable to our own Milky Way.

ALL HALE OBSERVATORIES EXCEPT SMALL MAGELLANIC CLOUD, NAVAL RESEARCH LABORATORY, AND NGC 7479, LICK OBSERVATORY

NGC 4486: brightest elliptical galaxy in the Virgo cluster.

NGC 147: elliptical galaxy in the constellation Cassiopeia.

NGC 205: elliptical galaxy in the constellation Andromeda.

Ellipticals contain a predominance of dying stars, spirals have a mixture of old and young stars, and irregulars consist mainly of young blue stars. The formation of stars in all of these galaxies apparently began at the same time — about 15 billion years ago. Since the galaxies do not differ in age, the ellipticals must have ceased new star formation soon after birth of the galaxy, while spirals and irregulars use their gas and dust to build stars more conservatively. Barred spirals, a subcategory, form arms at the ends of a barlike structure, instead of developing arms directly from the hub as regular spirals do.

NGC 2811: tightly wound spiral galaxy, in Hydra.

NGC 3031: bright spiral with well-defined outer arms, in Ursa Major.

NGC 628: loosely wound spiral with multiple arms, in Pisces.

Small Magellanic Cloud, one of two foggy star hordes near our own galaxy, classifies as a dwarf irregular galaxy. Crewmen on Ferdinand Magellan's ships sighted the Magellanic Clouds on their voyage around the world.

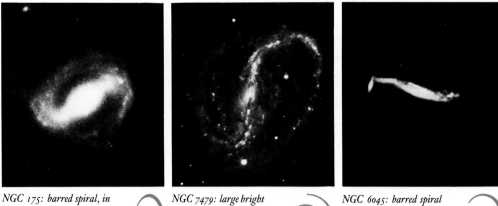

NGC 175: barred spiral, in the constellation Cetus, that has a cigar-shaped hub.

NGC 7479: large bright barred spiral with double arms, in Pegasus.

NGC 6045: barred spiral in the constellation named for the heroic Hercules.

strains our imagination to think of an entire galaxy exploding. Yet that is the startling picture that emerges from radio astronomy. Some radio galaxies pour out more energy from their exploded debris than all the luminous power of the Milky Way. The energy of a galactic explosion can exceed a billion trillion trillion trillion tons of TNT.

Radio astronomy is the key to much of our understanding of both normal and exploding galaxies. We can trace the beginning of this important branch of astronomy to an unpremeditated observation by a telephone company engineer in 1932; but it was anticipated as long ago as 1890 by Thomas A. Edison. The prolific inventor was the first to propose a radio antenna to pick up signals from the sun. He reasoned that if disturbances could be seen on the sun in visible light, they might also radiate radio waves. He knew of a huge field of iron ore in New Jersey and decided to put a loop of telephone wires around it. If the waves from the sun magnetized the ore, an electric current would be induced in the wire loop.

The poles for the wires were actually put in place, but the loop was never completed. It was just as well. The detector would not have been sensitive enough, and the long wavelengths to which the detector could respond would not have penetrated the atmosphere.

Appropriately enough, it was also in New Jersey that Karl Jansky, a Bell Laboratories engineer, undertook the tests that finally launched radio astronomy. To investigate interference in transoceanic radio-telephone communications, Jansky constructed a crude, 100-foot-diameter "merry-go-round" antenna and mounted it on four old wheels that rode on a cinder-block track.

The principal source of the cracks and bangs of radio static is thunderstorm activity. But Jansky found that even when there were no significant storms, he could still detect a persistent hiss. A careful observer, he noticed that the pattern of the static

roughly coincided with the sun's progress from east to west. But analysis of his records showed that the radio source appeared four minutes earlier each day, consistent with the movement of the stars. Finally Jansky determined the static came from the direction of the constellation Sagittarius; he correctly surmised that the noise came from the center of the Milky Way, which must be emitting radio waves as well as light.

Jansky's discovery aroused considerable popular interest. *The New York Times* of May 5, 1933, had headlines reading "New Radio Waves Traced to Center of Milky Way"; and the National Broadcasting Company played a brief recording of the hiss, which one reporter said sounded like steam escaping from a radiator.

Although Jansky published his findings, they received little attention from the scientific community. But they did interest a radio engineer, Grote Reber, who took up the investigation of galactic noise as a hobby. He spent about $2,000 to build a bowl-shaped antenna, 31 feet in diameter, in his backyard in Wheaton, Illinois. Since the dish had a hole in the center to allow water to drain, people suspected it had something to do with rainmaking. Imagine his neighbors' reaction when he explained that he would be searching for radio signals from the Milky Way!

A truly dedicated amateur, Reber for several years was the only radio astronomer in the world. After working a regular job during the day, he would sleep a few hours, then rouse himself at midnight, and from then until 6 a.m.—the period of least manmade interference—he would record radio waves in his basement laboratory.

Reber produced a rough map of the radio brightness of the sky, observing the strongest signals where the Milky Way star density was highest. Yet when he pointed his antenna at a specific bright star, he found no stellar radio emission, so he inferred what was later confirmed: The stars were not contributing significantly to the cosmic static;

therefore, the source must be the thin hydrogen gas of interstellar space.

Considering that he was working alone and in his spare time, Reber's achievement was truly remarkable. In 1940 he submitted his first scientific paper to *The Astrophysical Journal*. Editor Otto Struve could find no expert to comment on the paper, for it was the first work on radio astronomy ever offered to an astronomical publication. Struve himself went to Illinois to talk with Reber and to look at his telescope. Seeing was believing, and Struve accepted the paper.

THE YEARS of World War II, it now seems clear, represent a significant stage both in the evolution of radio astronomy and in the study of galaxies. Certainly that was not clear at the time; in general, little thought was given to either subject. But military need was a great spur to the development of radio technology; wartime access to new electronic devices helped the career of many an astronomer and technician; and new theories were being advanced that led in due time to the mapping of our galaxy.

Although we are part of the Milky Way, we have had almost as much difficulty studying our own galaxy as some of those far away. We see the Milky Way edge on, as a luminous ribbon of stars arching across the sky. The evidence tells us that the galaxy is disk-shaped and that our sun is located in its suburbs, about two-thirds of the distance from the center to the outskirts.

We get a much more comprehensive view of our neighbor Andromeda, two million light-years away and tilted about 15° from edge on. It is thought to be a near twin of the Milky Way. From our perspective we can see its beautiful spiral structure, an enormous pinwheel of stars laced with clouds of gas and dust.

Similar clouds cast a veil that hides the center of the Milky Way from the eyes of astronomers on earth. But within the last 20 years, radio, infrared, and X-ray astronomy have penetrated the gas and dust to the very core of the galaxy and found evidence that dispels earlier impressions of serene stability. The nucleus of our galaxy has been the site of violent explosions in the past, and may repeat the performance in the future. An enormous amount of matter is pouring out of the central region, and we have difficulty understanding its origin.

The earliest attempts to identify the spiral arms of the Milky Way focused on hot blue-white, supergiant stars. These are strung out along the arms like diamonds on a necklace.

Yet, bright as they are, we see few of them beyond 20,000 light-years from the sun, because of dust and haze. About the time optical astronomers were growing discouraged by the murkiness of the galactic disk, radio astronomy came along to help out.

In 1944, a Dutch student named Henk van de Hulst theorized that hydrogen atoms in the Milky Way generate detectable radio signals at a wavelength of 21.2 centimeters.

Atomic theory had already predicted that the electron in the hydrogen atom would occasionally — though very infrequently — flip over, reversing its spin about its axis; and that in the process the atom would release energy with a wavelength of 21 cm. But van de Hulst reasoned that, despite the infrequency of a single atom's flipping, the enormous number of hydrogen atoms available in the Milky Way should produce a measurable intensity of energy.

Because of wartime disruptions in communications, van de Hulst's prediction was not published widely. A similar prediction was made in the Soviet Union by Josef Shklovsky. Not until 1951 was there equipment available with sufficient sensitivity to detect the radio emission. It was first recorded by Harold Ewen and Edward Purcell at Harvard, and confirmed promptly by teams in the Netherlands and Australia. The discovery of signals from hydrogen gas in the Milky Way gave astronomers their first look at the most abundant form of galactic matter other than the stars themselves.

Not only does the hydrogen atom broadcast its presence, but it also tells us the direction of movement of the hydrogen clouds by either a lengthening (red shift) or a compressing (blue shift) of the wavelengths. Thus in the same way that we know galaxies are going away from us, we can tell whether the gas cloud is moving toward or away from us. In 21-cm hydrogen we have a highly useful tracer for mapping our galaxy.

The interstellar gas and dust permeating the space between the stars is 90 percent hydrogen. The gas is extremely dilute by earthly standards. As Lyman Spitzer expressed it, if one breath from a fly were spread throughout an evacuated space equal to the interior of the Empire State Building, the resulting density of gas would far exceed that of interstellar gas. But even though the gas is hundreds of millions of times thinner than the highest vacuum that can be created in the laboratory, it adds up to a substantial two percent of the mass of all the stars!

In our galaxy the gas is ten times more concentrated in the spiral arms than in between, and glows strongly in the 21-cm radiation. By measuring the velocity of the gas in the arms by means of 21-cm radiation, and collaborating with optical observers to obtain information on the galaxy's rotation, radio astronomers have been able to map the galaxy from the nucleus to its edge.

In the mapping process, they noted that hydrogen in the disk appears to flow outward from the core through the arms at speeds of 30 to 100 miles a second. Jan Oort and his Dutch colleagues believe that the flow of gas out of the nucleus and into the arms amounts to the equivalent of about one sun per year.

The nucleus of our galaxy is very complex, but radio, infrared, and X-ray astronomy are beginning to supply clues to explain the energy source for this enormous flow. Where does it all go? Guesses have included repeated explosions in the core, and gas circulation out through the arms and back—through a great spherical halo—into the nuclear region.

Close to the center of the galaxy lies a powerful radio source of synchrotron radiation, Sagittarius A. Since we know that explosive events—like the supernova that formed the Crab Nebula—create synchrotron radiation, we might suspect a similar explosion to have occurred in Sagittarius A.

Evidence for such an explosion about a million years ago comes from microwave studies of our galaxy's nucleus. Scientists have found interstellar molecules including those of water vapor, ammonia, carbon monoxide, formaldehyde, and ethyl alcohol. (One group of astronomers has expressed the total amount of alcohol in Sagittarius as equivalent to 40 thousand trillion trillion fifths at 100 proof.) These clouds of molecules girdle the nucleus in a giant smoke ring, about 2,000 light-years in diameter, steadily expanding outward at about 60 miles per second.

When observed in the infrared, the core of our galaxy resembles a Seyfert galaxy, although about 30,000 times less intense. Within a diameter of about three light-years, a hundred million suns brighten the sky.

George Rieke and Frank Low have found a few very bright sources concentrated near this region. Each of these sources radiates as much infrared energy as the total output of a million suns. The infrared may come from vast dust clouds that envelop the stars—but stars that must be a hundred times as massive as our sun. The infrared power from a narrow region near the core is nearly equal to the total visible light emitted by the entire galaxy.

THE ABILITY of a telescope—radio or optical—to distinguish a celestial object from other objects around it is called "resolving power," or resolution. It is related both to the diameter of the telescope and to the wavelengths the telescope is intended to receive. Radio waves are *(Continued on page 148)*

Shining with the light of myriad stars, the Milky Way stretches like a bright necklace across a painting of the night sky. Astronomer Bart J. Bok of the University of Arizona looks beyond the beauty of the Milky Way to its inner mysteries. Studying dark clouds of gas and dust that obscure the central regions, he isolated dense, almost circular patches called "globules." Some astronomers view these patches as potential nurseries for the birth of stars.

Many "nearby" stars speckle the painting. Just below our galaxy, to the right of center, float the two galaxies called the Small and Large Magellanic Clouds.

The Galactic Center

Hydrogen gas in interstellar space sends out a strong radio signal, Dutch physicist H. C. van de Hulst (right) correctly theorized. By tuning to its frequency, astronomers can map the shape of the Milky Way. Listening to this "song of hydrogen," astronomer S. Christian Simonson, III, of the University of Maryland produced the computer map below. Motions of hydrogen gas reveal star concentrations (black areas); thus the diagram suggests how a photograph of our galaxy would look. Through this method Dr. Simonson in 1975 located a neighboring dwarf galaxy at a distance of only 55,000 light-years. "I call this small star group 'Snickers,' for it's like a Milky Way, only peanuts," quips Simonson about its size.

Infrared astronomers such as Frank J. Low of the University of Arizona (extreme right) study the center of our galaxy as seen through its edge along the equatorial line (opposite, above). Dr. Low and colleague George H. Rieke cool their infrared detector with liquid helium; once it chills to near absolute zero (−273° C.), they survey the sky for temperature variations and make detailed maps, later color coded by hand. Red to yellow shows areas of greatest infrared emission; green to violet, areas of least emission.

JONATHAN S. BLAIR (ABOVE); S. CHRISTIAN SIMONSON,

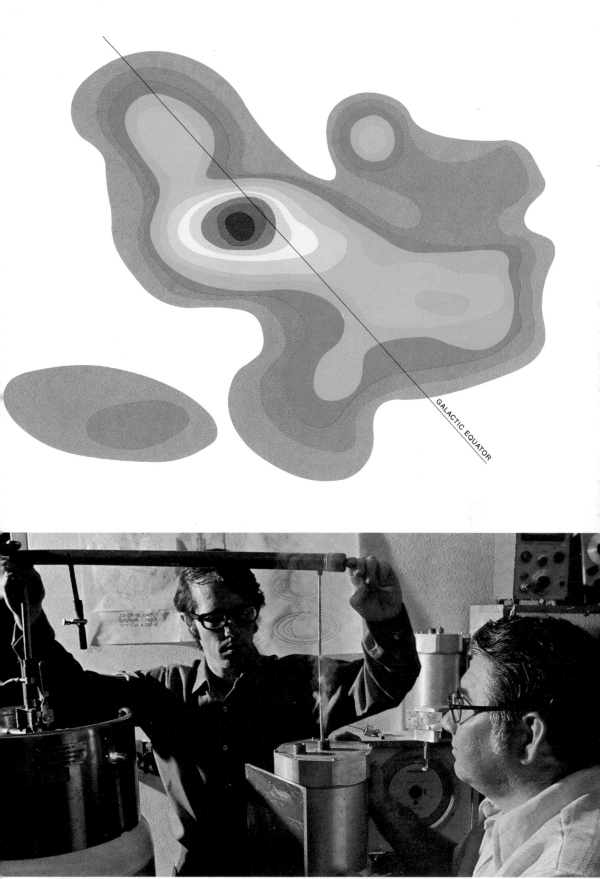

GALACTIC EQUATOR

Mile-long row of telescopes (opposite) at the Westerbork Observatory in the Netherlands investigates the structure of the universe. The 12 telescopes at Westerbork can measure as small an object as one big telescope a mile in diameter. A technician sits before the main console (above) operating the telescopes simply by pushing buttons.

Astronomer Jan H. Oort (top) pub-lished one of the first hydrogen maps of our galaxy, revealing multiple arms extending from a hub and showing that the Milky Way is spiral.

"You usually build an instrument for a very specific purpose, but the real advance of science lies just in the un-expected," explains Dr. Oort, here relaxing before a sundial in the shape of a radio telescope.

millions of times longer than light waves; thus, to obtain the same resolving power with a radio telescope as with a specific optical instrument, astronomers need a radio telescope millions of times larger.

Ideally, the 200-inch optical telescope at Palomar could distinguish a 25-cent piece at a hundred miles — or an object in the sky filling the arc of the same angle. Unfortunately, the shimmering air limits the resolving power of a good optical telescope; what we gain with very large size is not better resolution but more light-collecting power, which enables us to see to greater distances and collect light images in a shorter time.

For the surface of a mirror to reflect accurately, it must be smooth to within a few millionths of an inch. In contrast, a radio telescope for, say, meter wavelengths need be smooth only to within one-tenth the wavelength, or ten centimeters; it can even be constructed of wire mesh or punched with holes, and it will still seem smooth to radio waves. For shorter wavelengths, radio dishes have been built as large as a thousand feet in diameter with their surfaces shaped to a smoothness better than 1/8 inch.

To match the inherent resolving power of the 200-inch mirror, a radio telescope operating at one-meter wavelength would need to be as large in diameter as the earth. At any practical size, a single-dish radio telescope would be unable to identify starlike radio sources with specific optical images.

Fortunately, there is a solution: Combinations of radio telescopes can be used in unison as "interferometers." The receivers of two or more instruments are coupled by electrical cables, and their signals are then fed to a central processor. As the earth rotates, the radio waves arriving at a pair of dishes will alternately reinforce and cancel each other, thus producing "interference fringes." Because these fringes are closely spaced, the interferometer can measure very small sources of radio emission; the closer the fringes, the more precise the measure. The resolving power of two antennas a mile

apart is equal to that of a single dish a mile in diameter, although the amount of energy collected by the smaller dishes is proportionately less.

The principle of improving resolution by wide separation can be carried to distances of thousands of miles, up to the diameter of the earth. Observations using this maximum separation could distinguish a source the size of a man standing on the moon. But of course the connecting cables would become of impractical length. The answer has been to eliminate the cables and use precise atomic clocks in timing and recording signals at each telescope site. Such clocks are so accurate they would not lose a second in a million years. The record tapes are then matched by their clock references to give the interference patterns.

Radio sources have been observed with telescopes sitting as far apart as Russia and the United States. The longest base line used so far runs from the Algonquin Observatory in Canada to Parkes Observatory in Australia, a distance of 6,700 miles — not much short of the 7,900-mile diameter of the earth.

Ken Kellermann of the National Radio Astronomy Observatory at Green Bank, West Virginia, tells of setting up, in 1969, such an experiment — called Very Long Base-line Interferometry — with Russian scientists at the Crimean Observatory on the shores of the Black Sea. Synchronizing the clock signals was of paramount importance. The battery of the first atomic clock carried by scientists from the United States failed en route to the Crimea, and the clock stopped. A second clock, set by Swedish radio astronomers in Stockholm, was then flown directly to Leningrad where it was collected by Kellermann and transported to the Crimea by plane and car while its battery rapidly weakened.

Passage through airports and customs gates with a large piece of electronic equipment turned up innumerable problems. Meanwhile Kellermann had to keep an eye out for opportunities to plug the clock into a power outlet. It was nip and tuck, but he kept the clock's atomic heart beating all the way to its destination.

After completing the observations—which involved mastering telephone communications between West Virginia and the Crimea by way of Moscow—all that remained was to get the tapes back to Green Bank. That required convincing customs officials of both countries that the tapes contained nothing more than cosmic radio noise! But in the end the scientific objectives were met, and the results were jointly published under the names of the Russian and American teams.

IN 1942 BRITISH RADAR detected intense static on such a scale that it could not be attributed to local sources of interference, and it appeared that the Germans had come up with a new and extremely powerful jamming device. When Stanley Hey analyzed reports from several sites, however, he came to the conclusion that the signals were produced not by Nazi jammers but by the sun. The static rose and fell with the rising and setting sun, and eventually disappeared when a large group of sunspots passed to the back side of the sphere.

Stimulated by their wartime experiences with radio and radar, a considerable number of scientists like Hey later turned to radio astronomy. John Bolton's description of the discovery of the radio source Taurus A in the Crab Nebula reflects their pioneering persistence:

"Gordon Stanley and I discovered Taurus A in November 1947. It took six months to construct equipment capable of observing it each day with some certainty, then three months of observing from an ex-Army gunlaying trailer at two sites in New Zealand, and a further three months manually reducing records and making calculations to show that Taurus A was very probably the Crab Nebula.

"Nowadays, one can determine much more accurate positions of radio sources 100,000 times fainter than the Crab Nebula in something like 1/30,000 of the time without leaving the warmth of the telescope control room and without setting pen to paper."

Immediately after the war, Hey began to use the receiving portion of his radar to observe the sky, and soon made a major discovery. In the direction of Cygnus he detected strong, fluctuating radio signals, and decided that the source must be starlike. But astronomers were hardly prepared for the revelation of its true nature. In the summer of 1951, F. Graham Smith of Cambridge University mounted two war-surplus radar dishes a thousand feet apart and linked their receivers with cables. The separation was measured to an accuracy of one inch; and Smith was able to use this interferometer to locate the radio source called Cygnus A so precisely that Walter Baade and Rudolph Minkowski could search for visual evidence of the source with Palomar's 200-inch telescope. In the stipulated area they found a very faint nebulosity. Its diffuseness suggested that it was a galaxy, but its shape was peculiar: Two irregular condensations seemed to be joined, surrounded by an extended faint halo.

The double structure strongly suggested two galaxies in collision. From the red shift, the distance was found to be 700 million light-years. The startling conclusion was that, at such a great distance, Cygnus A had to be radiating 70 billion trillion trillion kilowatts of radio energy, more than three times the luminous power and three million times the radio power of the entire Milky Way! It would take ten billion Crab Nebulae to match the radio emission of Cygnus A.

For many years thereafter Baade persisted in believing, despite the skepticism of several prominent astronomers, that the only way such intense radio waves could be generated was by the collision of galaxies; and he offered to bet Minkowski, one of the doubters, a bottle of whiskey. For a while

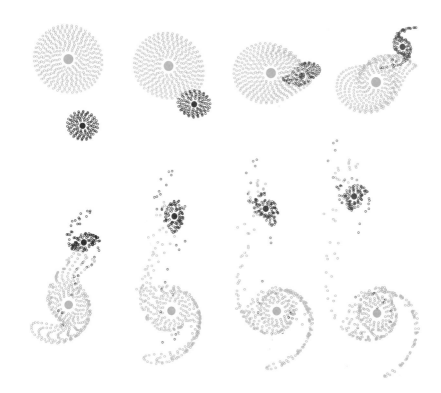

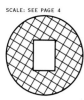
SCALE: SEE PAGE 4

Gravitational pull between orbiting galaxies distorts their shapes, according to a theory demonstrated by computer diagrams (above and below). Spinning counterclockwise, the two galaxies (top row) pass near each other. Streamers of stellar matter break away (next row), pulled by gravity like tides on earth. Testing this theory on the Whirlpool galaxy (M 51, opposite), the computer produced the diagrams below. Seen face on (below, left) the Whirlpool's predicted tidal arm (in blue) appears to connect the two galaxies. A side view (below, right) reveals their true separation. The pericenter represents the point of closest approach between the two, and the cross indicates the center of gravity of the system. Two mathematician-astrophysicist brothers, Alar and Juri Toomre, developed this computer model to support their own theory of galactic tidal waves.

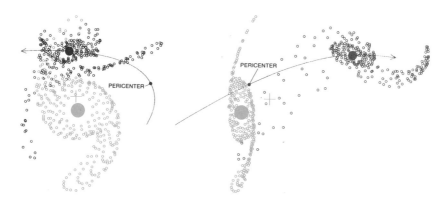

it appeared that Baade was right, and Minkowski delivered the prize—a hip flask. But two days later, Baade reported, Minkowski visited his home and drank the whole bottle.

Meanwhile, the controversy drew the attention of the press, and at least one British broadcast, inviting its listeners to hear "the thunder of colliding galaxies," played a recording of the signals from Cygnus A.

But the image of two colliding galaxies was deceptive. We know now that it would be impossible to generate the enormous radio power of Cygnus A even from a collision on a galactic scale. For all the colossal mass of stars and dust, colliding galaxies are so dispersed that they brush through each other without catastrophic impact. We now believe that Cygnus A is the product of a gigantic explosion at the center of the visible galaxy. Present-day radio telescopes reveal two large clouds of energetic gas, or plasma, that have been shot out in opposite directions to distances of a hundred thousand light-years. Each cloud is a magnetic bag laced with cosmic-ray electrons and protons equal in energy to the mass of about a million suns. Yet the volume is so large that the gas density is billions of times less than the best laboratory vacuum.

Several other exploding galaxies have recently been given careful study. The optical galaxy NGC 5128 is located at the center of the extended double radio source Centaurus A. Its strong radio sources are centered on a pair of dense, radio-emitting clouds a million light-years apart. But within the border of the optical galaxy a rather recent explosion produced a pair of clouds only 20,000 light-years apart. At the very center of the galaxy, a bright compact radio and infrared source only 12 light-days in diameter is furiously active.

Some of the first test photographs by John Graham with the new 158-inch reflector at Cerro Tololo in Chile have shown irregular knotted filaments jetting out 130,000 light-years from the center of the galaxy. Loose aggregates of young blue supergiant stars are strung along the filaments. Graham and his colleagues believe these new photographs suggest the formation of entire satellite systems of stars from gas and dust that have exploded out of the nucleus.

In the center of the Virgo cluster at a distance of 35 million light-years is a giant elliptical galaxy, M 87, also known to the radio astronomers as Virgo A. Thirty times as massive as the Milky Way, it is speckled with more than a thousand globular clusters of stars. Protruding from its core to a distance of 4,000 light-years is a bright blue jet of gas dotted with dense knots. Observers have found a powerful radio source concentrated in the nucleus, which is no more than two light-months in diameter. Intense X-ray emission also seems to originate there.

Roger Lynds has photographed still another exploding galaxy, NGC 1275, or Perseus A, in the red light of the hydrogen atom. It bears a striking resemblance to the Crab Nebula, but the scale is 10,000 times as great. From its nucleus, filaments of gas stretch outward tens of thousands of light-years. At the center is a compact radio source about one light-year in diameter. Explosive radio outbursts occur about every ten years, and the debris streams outward to form an extended halo. The galaxy appears to have been blasting away for several million years in this fashion. In the infrared range the nucleus is far, far brighter than in the visible, and its X-ray power almost matches the infrared radiation.

But even exploding galaxies seem almost calm when compared with a newly discovered cosmic powerhouse—the quasar.

THE DISCOVERY OF QUASARS—perhaps the brightest objects in the universe—is a classic example of the interplay of the old and new astronomies. What we know as quasars had appeared on photographs taken with the large optical telescopes for many years, but were undistinguished in any obvious way from faint stars in our own galaxy. By 1960 the growing activity of radio astronomy had detected several hundred radio sources. Some coincided with well known optical objects, including nebulae and galaxies; others were visible or invisible pointlike sources that scintillated in radio brightness the way optical stars twinkle, and these came to be called "radio stars."

At any reasonable assumed distances in the galaxy, these objects were enormously more powerful radio generators than the sun. It was difficult for astronomers to comprehend how an ordinary star could produce such strong signals.

The earliest observations with radio interferometers confirmed that the so-called radio stars were, indeed, very much smaller than radio galaxies and nebulae. At the Owens Valley Radio Astronomy Observatory of the California Institute of Technology, John Bolton and Thomas Matthews identified the radio source 3C 48 with a bluish star. In the customary follow-up, Palomar astronomer Allan Sandage measured the star's brightness, and his colleague Jesse Greenstein obtained its spectrum—but they were completely perplexed by the pattern of spectral lines, which were very broad and corresponded to those of no known elements. It was absurd to assume that some new element was responsible. Even with the 200-inch telescope, 3C 48 retained a starlike appearance, and its color most nearly resembled that of a white dwarf. Most astronomers believed that it was indeed a nearby star, perhaps a few hundred light-years distant. Supporting these views was evidence of large variations in brightness in periods of less than a day—indicating that the size of the varying object could not exceed one light-day.

In late 1962 a British astronomer, Cyril Hazard, pointed the 210-foot radio telescope of Parkes Observatory in New South Wales toward 3C 273, a radio source in the constellation Virgo that lay along the path traveled by the moon. The obscuring passage of the moon across the radio source could be used to study it in the same manner as the moon had been used to study eclipses of the sun.

Hazard observed the radio signal while he waited for the moon to pass in front of the source. At the expected moment the signal disappeared. Hazard waited patiently for the signal from 3C 273 to resume; but when it did, he was surprised that its intensity measured only half strength for about ten seconds before suddenly recovering its full value. Mystery object 3C 273 had gone into eclipse as a single source and emerged as two sources.

Geometry gave the explanation. The edge of the moon is, of course, an arc. If the radio source were a double object and both components happened to align with the leading edge of the moon, they would disappear simultaneously; but on emergence, with the other edge of the moon curving in the opposite direction, each component would reappear separately.

Succeeding checks in the next few months enabled Hazard to associate the radio sources in 3C 273 with a faint blue starlike object that had a strange protrusion on one side like a luminous jet of gas. A star with a jet! No such star had ever been seen before. One radio component was centered on the star, and the other fit the tip of the jet. Could 3C 273 actually be a galaxy like M 87, with its exploded tail?

The precise position was communicated to Palomar, where a young Dutch astronomer, Maarten Schmidt, went to work to obtain the spectrum of the starlike component.

When Schmidt tried to analyze the spectrum of 3C 273, he was baffled. After weeks of deep puzzlement, it suddenly occurred to him that the spacing of the spectral lines resembled the simple pattern of hydrogen emission, except that the entire pattern was displaced extremely far down toward the red end of the spectrum. "That night," Schmidt recalls, "I went home in a state of disbelief. I said to *(Continued on page 158)*

NATIONAL GEOGRAPHIC PHOTOGRAPHER JAMES P. BLAIR (OPPOSITE); HALE OBSERVATORIES

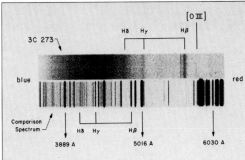

Mystery of the Quasars

Astronomical enigma, quasars continue to puzzle skywatchers. After their discovery in 1960, scientists found that the spectral patterns of these faint starlike objects did not match the standard comparison spectrum that identifies chemical elements in the stars. Then, in 1963, Maarten Schmidt of the Hale Observatories (opposite) suddenly realized that the familiar pattern existed, but in an unexpected place on the spectrum. Examining quasar 3C 273 (above), he noticed that lines of the spectrum representing hydrogen (bracketed areas in diagram) had shifted radically toward the red end.

Assuming that the red shift meant that the source of light was moving away from the observer, Schmidt calculated 3C 273 as some two billion light-years distant. Red shifts have since been measured for hundreds of quasars. Some have very large red shifts that indicate an incredible speed of more than 90 percent that of light. These quasars may lie so far away that it took their light three-quarters of the age of the universe to reach us. At such distances quasars must shine with the brightness of a hundred galaxies to be observable at all. Some astronomers question that quasars are very distant, but they have produced no acceptable explanation for the red shift.

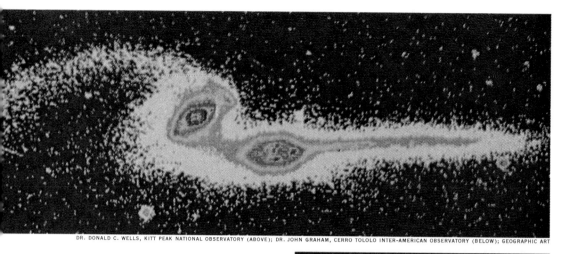

DR. DONALD C. WELLS, KITT PEAK NATIONAL OBSERVATORY (ABOVE); DR. JOHN GRAHAM, CERRO TOLOLO INTER-AMERICAN OBSERVATORY (BELOW); GEOGRAPHIC ART

Mapping Galaxies

Near collision: Two galaxies (top), called "The Mice," appear to have brushed past each other, leaving tails of gas, dust, and stars. Gravity pulls out a bridge of hydrogen between the pair. The color-coded photograph displays degrees of brightness, with greatest intensity at the center, least at the outer edges.

Another galaxy, Centaurus A (right and rectangle in diagram below), gives off one of the strongest radio signals in the sky. A dark band of gas and dust obscures its center. When mapped, the area of Centaurus A's radio source appears a hundred times larger than its optical size (comparison below).

The painting opposite shows how two telescopes in different parts of the world provide detailed information about radiation sources. Here, both receive signals from the radio galaxy Cygnus A. Arrows define the limits of the radio source as enlarged by the map made by the two telescopes. The white area at the center represents the visible galaxy that emitted the pattern of radio waves.

157

my wife, 'Something really incredible happened to me today.' "

The fit was too precise to be mere chance, but the implications were astonishing: If the red shift meant in this case what astronomers had come to accept, then 3C 273 was two billion light-years away, rushing into space at 28,400 miles per second. The "radio star" had to be far out among the galaxies. But to appear as bright as it did from that distance, the amount of energy the object radiated must be equal to 200 ordinary galaxies. When Schmidt passed this startling news to Jesse Greenstein, his colleague immediately checked the object known as 3C 48. Again the lines fell into the pattern of red-shifted hydrogen, but in this case the red-shift distance was even more difficult to believe: 3.6 billion light-years!

The enormous luminosity of 3C 273, if the red-shift distance was correct, meant it had to have the mass of a galaxy even though it appeared far smaller than a typical galaxy. In the uncertainty of how to categorize it, astronomers used the term "quasi-stellar" object or source; and Hong-Yee Chiu, a Shanghai-born American astronomer, coined the short form "quasar."

The number of known quasars now runs into the hundreds, and the largest red shifts correspond to velocities of 91 percent of the speed of light, in contrast to about 46 percent for the most distant known galaxy. Most astronomers believe the red shift is associated with Hubble's expanding universe; but then it becomes very difficult to explain the quasar's enormous radiant power.

The quasar 3C 273 radiates a hundred times more light than the brightest ordinary galaxy. It produces ten times as much X-ray power as light, and its infrared radiation is a hundred thousand to a million times the entire optical power of our own galaxy. All of this energy is released from a volume less than a light-year in diameter—whereas the Milky Way is 100,000 light-years across. If the quasar were a short-lived explosion like a supernova, the brightness might be under-

standable. But 3C 273 appears approximately the same on photographic plates taken 85 years ago, whereas a supernova typically fades in a few weeks.

If 3C 273 is at the distance implied by the red shift, then its jet measures 150,000 light-years. So even if it grew with the speed of light, it must be at least 150,000 years old.

Astronomers have long believed that objects as large as galaxies could not change rapidly enough for us to detect results within the few thousand years of historic time. For if an average galaxy measures 50,000 light-years across, any explosion, even if it spread with the speed of light, would take 50,000 years to illuminate the entire galaxy. The change in overall brightness would then take a comparable period for us to detect.

For almost 70 years—from 1885 to 1954 —the Harvard College Observatory systematically photographed the sky. In 1970 it resumed its record-keeping. In the last few years William Liller and his associates have been searching this library of half a million photographs for quasars.

They have found surprising evidence of variations. The quasar 3C 273, for example, has fluctuated slightly in brightness from year to year. Yet it seems incredible that it could be small enough to make such a rapid change possible.

And if that possibility seems shocking, how about even faster changes? In April 1937—the photographic plates revealed— another quasar, 3C 279, flared brilliantly to become intrinsically the most powerful light in the sky, although it was so far away it remained invisible to the unaided eye. Over a span of 13 days its intensity was the equivalent of suddenly switching on tens of thousands of galaxies like the Milky Way. Since the flare was so short-lived, the size of the source could not have exceeded a few light-weeks—seemingly an impossibility.

Now, through interferometry, the quasar 3C 273 has been resolved into five parts. Close to the center of the optical source lie two compact radio components, each less

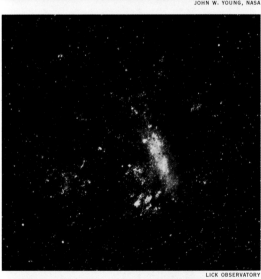

From the moon, a camera transported by Apollo 16 (above) captured ultraviolet images of the Large Magellanic Cloud (lower left). The ultraviolet radiation emitted by hot young stars reveals a very different view of the galaxy from that recorded at the same scale by an earth-based optical telescope (left). Developer of the ultraviolet camera, George R. Carruthers (below) of the Naval Research Laboratory in Washington, D.C., holds components of a new ultraviolet camera proposed for Space Shuttle flights. In the foreground rests a smaller camera of a type used on Apollo 16.

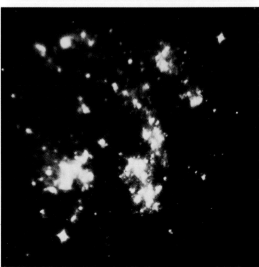

JOHN LUTNES, KITT PEAK NATIONAL OBSERVATORY

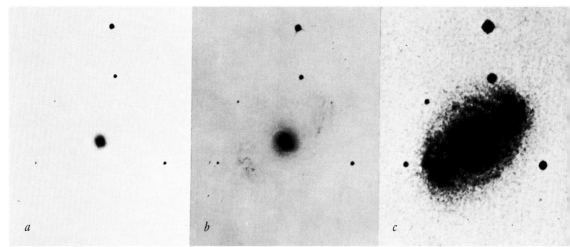

a   *b*   *c*

YERKES OBSERVATORY FROM MOUNT WILSON OBSERVATORY PLATE (LEFT AND CENTER); GLASS COPY OF NATIONAL GEOGRAPHIC SOCIETY–PALOMAR OBSERVATORY SKY SURVEY

Quasars—Near or Far?

Focusing a newly developed camera attached to an optical telescope at Kitt Peak, C. Roger Lynds (opposite) prepares to take a short exposure to show the surface of a star. Dr. Lynds has perfected the use of electronic cameras for photographing quasars and other faint galaxies.

If quasars exist as far away as their red shifts indicate, they must produce enormous amounts of visible and radio energy—quantities that scientists cannot account for. Russian astrophysicist Josef Shklovsky (above, left) first recognized that high-energy electrons spiraling in magnetic fields produce powerful radio emissions associated with supernovae. Such emissions, called synchrotron radiation, vary greatly in intensity when recorded from radio galaxies, quasars, and even pulsars. "Quasars are not as distant as their red shifts indicate," argues Halton Arp of Hale Observatories, here scanning a photographic plate of a field of galaxies (above, right). "Therefore they do not have to release such unprecedented amounts of energy. Their red shifts, however, would have to be caused by some mechanism not currently recognized in physics." For evidence he has found numerous quasars apparently associated with relatively nearby galaxies.

Neither ordinary star nor simple galaxy, a Seyfert galaxy, NGC 4151 (opposite), appears quasar-like in a short exposure (a). A longer exposure (b) begins to reveal a spiral structure. Only a long exposure (c) clearly reveals the dim spiral arms of the Seyfert galaxy. Such galaxies may fill in the evolutionary gap between quasars and ordinary galaxies.

than a tenth of a light-year in diameter, and separated by only a third of a light-year. Two others are strung out about two light-years and 15 light-years from the center. Out at the tip of the jet lies the largest component, about 200 light-years in diameter.

As radio telescopes push the limits downward on the size of quasars, the mystery deepens. Even if all its mass could be converted to radiation, a quasar would need the mass of at least a hundred million suns to produce so much energy—and this enormous amount of material would be crowded into a volume possibly less than a tenth of a light-year in diameter.

Philip Morrison of the Massachusetts Institute of Technology, pointing out the extreme inefficiency of converting energy from nuclear fusion into synchrotron radiation, suggests an answer to the puzzle: that the quasar resembles a pulsar on a gigantic scale. One of the amazing aspects of a pulsar is that it converts its energy of rotation into cosmic rays and synchrotron radiation with very high efficiency. Morrison calls his spinning ball of a hundred million solar masses a "spinar." Given a sufficiently high magnetic field and a rotation period of about a year, the pulsar mechanism theoretically should work. As the quasar loses energy, it contracts and spins faster. The faster it spins, the brighter it becomes.

To support his theory, Morrison quotes evidence of variations in the quasar 3C 345 that resemble pulsations about a year apart. Each surge contains the power of a trillion suns. His model is attractive, but we need more observations to feel convinced.

If quasars are extremely condensed, yet contain stellar mass comparable to an ordinary galaxy, the traffic must be like a freeway at rush hour. In the vicinity of our sun, by contrast, two stars are unlikely to collide in the entire life of the galaxy. Sir Arthur Eddington described the openness of galactic space very well: "Imagine thirty cricket balls roaming the whole interior of the earth; the stars roaming the heavens are just as little crowded and run as little risk of collision as the cricket balls."

But within a quasar, stars may collide as often as once a day. Each collision could release a huge burst of energy to enhance the normal light of the stars. Some collisions may combine stars and speed them to the supernova state. A rapid succession of supernovae would greatly brighten the quasar. Furthermore, any pulsars formed in the supernova process would add their powerful outputs. One pulsar formed per day could account for all the power of 3C 273.

In many respects the Seyfert galaxies— the special-case spirals with extremely high output of energy—have properties intermediate between quasars and ordinary galaxies. Could quasars, Seyferts, radio galaxies, and ordinary galaxies form an evolutionary chain? Could a quasar be an earlier stage of the nucleus of a Seyfert galaxy? The evidence is still very incomplete.

Possibly all the speculations we have described about energy sources are wrong. A radically different view was proposed several years ago by Soviet astrophysicist I. D. Novikov and Israeli physicist Yuval Ne'eman. They believe that the nuclei of quasars may have been there from the beginning—that galaxies grow out of their nuclei. The cores may be fragments of primordial material exploded by the big bang, or material exploding back into the universe after a massive collapse. Each galactic nucleus has become the site of a "little big bang."

No theory yet put forward has enough evidence to support it. The number of these theories only demonstrates the groping of scientists for some explanation of these fantastic phenomena.

So far, nature is too clever for us.

Neighbor in space, the spiral-shaped Andromeda galaxy (M 31) resembles the Milky Way. Two small satellite galaxies tag alongside Andromeda, whose center glows primarily with the light of old giant stars.

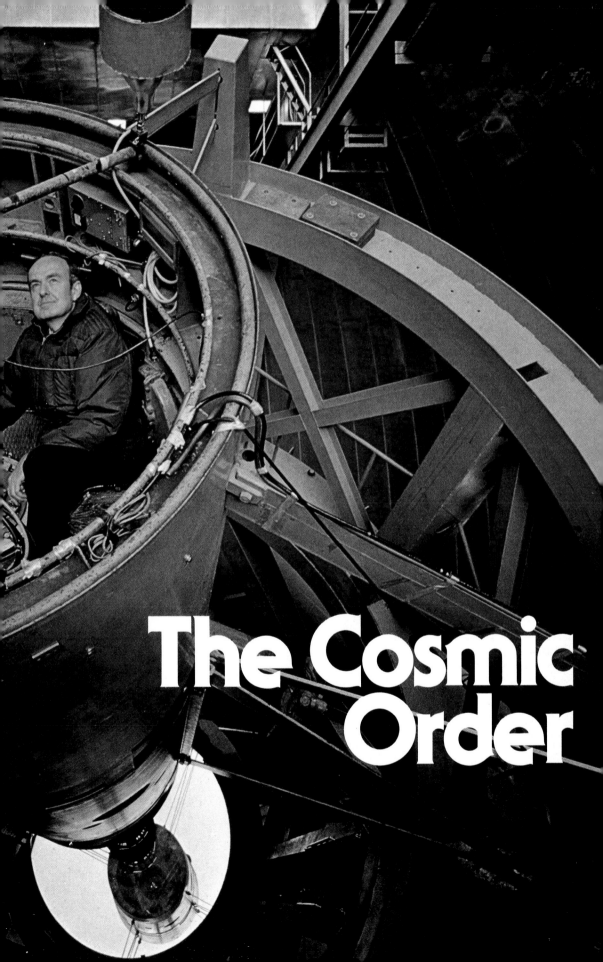

The Cosmic
Order

the theory of explosive creation or the big bang, the universe is the glowing debris of a huge fireball that exploded about 16 billion years ago. If so, did the explosion have such force that everything will eventually scatter to infinity? Or is the universe, as still other astronomers believe, oscillating between explosion and collapse? If that concept is correct, pieces of the universe will eventually fall back together in a fiery collapse that will provide the conditions necessary for an explosive rebirth.

All three theories, of course, may prove wrong; still others may evolve as the universe poses more and more questions. Astronomers do not have the answers, but in recent years they have made a flood of amazing discoveries that have led to a better understanding of what goes on in the universe. We now know that we are bathed in radiation that appears to be left over from the blast of creation—the echo of the big bang. Distant galaxies and quasars broadcast radio signals that give clues to the earliest history of the coming together of matter into groups of stars. From X-ray astronomy, we find hints of invisible matter that may outweigh all the visible material in the universe.

Each new observation should bring us closer to the truth. Often, however, the evidence is conflicting and the truth is elusive. Astronomers do not always interpret what they see the same way. But this is the way of science: resolving the conflicts, matching theories and observations, working step by step to solve the mysteries of the universe. As Philip Morrison once said: Two astronomers may disagree on their results, but finally the facts—and not another person—will be the arbiter.

Perhaps the truth will be impossible to find; but the study of the origin and evolution of the universe is one of the greatest adventures of the mind.

Why is the sky dark at night? This question, almost childlike, yet important, baffled Edmund Halley in 1720. Heinrich Olbers still sought the answer a hundred

IT IS IMPOSSIBLE for any sensitive person to look at a star-filled sky without being stirred by thoughts of creation and eternity. The mystery of the origin and destiny of the universe haunts us throughout our lives.

Philosophically, there can be just as many versions of how the universe began and how it will end as there are philosophers. Nobel laureate Bertrand Russell once noted that we have no way of proving that the universe was not created five minutes ago complete with its apparent history and all our memories.

The universe of the scientist, however, must fit with what he sees and measures with his telescopes and other instruments. His story of what has happened must be consistent with the prevailing knowledge of physics, chemistry, and other sciences. The field of study that attempts to combine this body of scientific knowledge into an explanation of the origin and structure of the universe is called cosmology.

The choices of cosmological theories are few, but they are vastly different. Astronomers who believe in the steady-state theory contend the universe has always existed and will continue to survive without noticeable change. To those who support

Overleaf: In an observer's cage of the Hale Telescope, Allan Sandage prepares cameras for photographing distant galaxies. Analysis of their light gives astronomers clues to the birth and fate of our universe.
NATIONAL GEOGRAPHIC PHOTOGRAPHER JAMES P. BLAIR

years later. The fact that the sky is dark turns out to be a significant clue to the nature of the universe.

Skywatchers in Olbers' time found increasingly more stars as larger telescopes made it possible for them to see fainter stars deeper in space. Olbers reasoned: Suppose the universe is unchanging, uniform, and infinite. Stars beyond stars would fill all space. In every direction we looked, our gaze would eventually find a star, thus creating a solid blanket of stars from the billions of pinpoints of light. Day and night the entire sky would be ablaze with light. This does not happen, of course, and the puzzle came to be known as Olbers' Paradox.

Olbers realized that one answer to the riddle could be that stars do not extend infinitely into space. This meant that the universe had limits. But Olbers believed this idea inconsistent with the infinite power of God. He tried to resolve his difficulty by assuming the existence of enough obscuring gas and dust in space to absorb the light of the distant stars and to produce the darkness of night.

We now know that interstellar matter does obscure stars. But when starlight is absorbed, the temperature of the matter rises. In time, a cloud of gas and dust would get hotter and hotter until, finally, it would radiate as much light as it received. In an infinite universe the cloud ultimately would become as bright as the stars, and we would have light constantly. Thus the darkness of the night sky proves that the universe is limited.

The answers to Olbers' Paradox are apparent if we concede that the universe is changing. In an expanding universe most of the light from the farthest, fastest-moving galaxies is red-shifted out of visible range. Also, the most distant galaxies are so far from us and are rushing away so fast that their light has not had time to catch up with us. Even if the universe were static and infinite, it would take ten trillion trillion years for the light of the sky to become as bright as the surface of the sun. Obviously this would be impossible because most stars shine for only a few billion years.

We have solved Olbers' riddle—the sky is dark because the universe has limits and is expanding. However, today we know the heavens are dark only to our human eyes. If we use the "eyes" of instruments that detect radio waves, infrared radiation, X rays, and gamma rays, we realize that the sky is filled with background radiation. In 1965—33 years after Karl Jansky of Bell Laboratories accidentally discovered radio waves coming from space—two other Bell researchers, again purely by chance, detected what may be the earliest light of the universe.

At Holmdel, New Jersey, Arno Penzias and Robert Wilson were using a supersensitive, 20-foot horn-shaped antenna to listen in on the sounds of the universe. The antenna originally had been built by Bell engineers to detect radio waves bounced off Echo balloon satellites. Penzias and Wilson were trying to eliminate all interference from their receiver so that they could measure the faint radio waves from the Milky Way. They succeeded in removing the effects of aircraft radar, ground radar, and noise from radio broadcasting stations. But the heat in their radio receiver itself caused troublesome interference. They suppressed this by cooling the receiver with liquid helium to $-269°$ C., only 4 degrees above absolute zero—the temperature at which all motion in atoms and molecules stops.

They still found a residual hiss that was a hundred times more intense than they had expected. Furthermore, the noise was evenly spread over the sky and was present day and night.

At about the same time, Robert H. Dicke and his colleagues at Princeton, just 40 miles from Holmdel, were preparing to search for the same kind of signals from space. Dicke reasoned that the big bang not only must have scattered matter that condensed into galaxies, but also must have released a tremendous blast of radiation. He calculated that the expansion of the universe

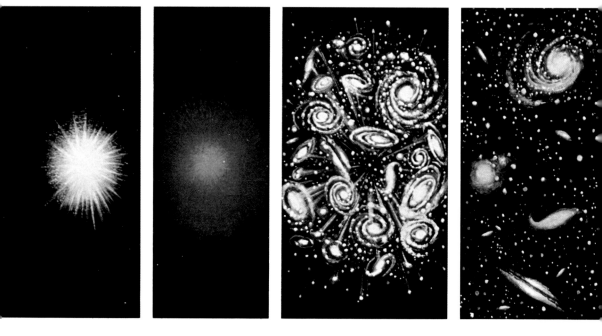

COURTESY OF BELL LABORATORIES (BELOW); J. R. EYERMAN (RIGHT)

PAINTINGS BY DAVIS MELTZER

 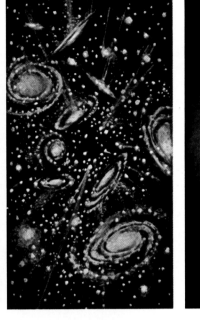 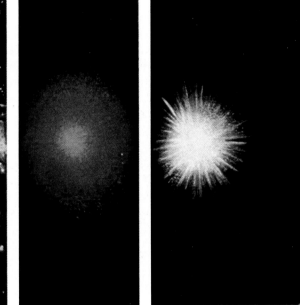

Cosmic theories unfold in a conception of the distant past and far future of the universe as seen by an observer outside the universe. According to the theory of the "big bang," a white-hot, superdense mass explodes (first panel). As it expands, the fireball cools; gas clouds form and begin to condense into stars. Galaxies rush outward into space (third) and may rush on forever. The oscillating-universe theory assumes, however, that gravitational force will stop the expansion after about 40 billion years (fourth). Then begins a cosmic rush inward (fifth). Eighty billion years after its birth, the universe falls back on itself (sixth). Under this oscillating or "repeated-big-bang" theory, another explosion eventually occurs (last panel) and forms a new universe.

Scientists think our universe is expanding from an explosion that occurred 16 billion years ago. The light from that blast should still exist, but in a different form, a "fossil remnant" of the big bang. In 1965, working with a hornlike antenna (far left), researchers at the Bell Laboratories in Holmdel, New Jersey, detected the predicted remnant in the radio wave area of the spectrum. Several physicists at Princeton University, under the leadership of Robert H. Dicke (left), predicted the existence of the radiation and confirmed the Holmdel finding. Physicists and astronomers continue to probe this ancient message of creation.

would cool the radiation to several degrees above absolute zero.

When James Peebles, another Princeton researcher, prepared a paper on these ideas, Penzias and Wilson realized the significance of the residual noise that they could not eliminate from their receiver. It was a dramatic breakthrough in cosmology.

The radio whisper was the ancient message of creation – the flash of the early blazing universe, long before the first galaxies were born. Over some 16 billion years this light had been red-shifted and stretched to radio wavelengths that would come from a source that had a temperature of only 3° above absolute zero. The radiation was easily detectable by the cool receiver used by the Bell scientists.

Penzias and Wilson later published the results of their work in a scientific journal. And in the same issue, Dicke and his colleagues gave their interpretation of what had been heard, calling it "primordial cosmic fireball radiation."

Physicists then recalled that George Gamow had predicted the existence of just such a microwave background radiation as early as 1948. He and his colleagues, Ralph Alpher and Robert Herman, had concluded that the radiation of the primordial fireball should still be detectable today, but would be tremendously red-shifted. Gamow never conducted a search for the radiation, and radio astronomers were then preoccupied with their first surveys of the sky and not yet seriously concerned with cosmology.

Apparently Dicke himself almost stumbled on the discovery as far back as 1945 when he was studying radio emission from the atmosphere. He had remarked that if there were any sky radiation from outside the atmosphere, it would have to be less than 20° above absolute zero. Gamow was not aware of Dicke's work, and Dicke missed the significance of Gamow's paper. By 1964, when Dicke had begun searching for cosmological background radiation, he had even forgotten his own prediction.

"WHAT WAS GOD doing before He created Heaven and Earth?" St. Augustine once asked, according to legend. He answered his own question: "He was preparing a Hell for those who inquire into such high matters." Those who have inquired into such matters can never know for sure how the fireball came to be, but once its existence is accepted, scientists can piece together a theoretical history of the universe from the explosion of creation to the present.

The story starts with the explosion of a superdense, superhot fireball. At the instant the fireball exploded, it enclosed almost as much space as that encompassed by the orbit of earth around the sun. Packed inside was all the material in today's universe. Its temperature exceeded a hundred trillion degrees – an inferno so hot that nuclear reactions created every conceivable type of particle that goes into the makeup of atoms. The exploding fireball expanded rapidly as particles destroyed one another, releasing every form of high energy radiation. The universe was launched on its expansion.

In only a few minutes, the maelstrom of particles and radiation cooled below a billion degrees. The cooling slowed down the nuclear furnace, and matter became more stable. From the holocaust emerged an electrified gas made up of atomic particles and packets of radiant energy – the raw material of the present universe. For a hundred thousand years, matter was ionized by the fireball of radiation. When temperatures dropped to about 5,000°, the radiation buffeting slowed down and electrons and nuclei joined to form atoms. The expanded fireball was then about a thousandth the size of the present universe, and its surface was as brilliant as the sun.

The young universe continued to grow. As tens of millions of years passed, particles clumped into clouds and, eventually, formed galaxies of stars. From then until now, both the universe and the wavelengths of photons stretched a thousandfold. The stretched radiation from the fireball is the hiss the

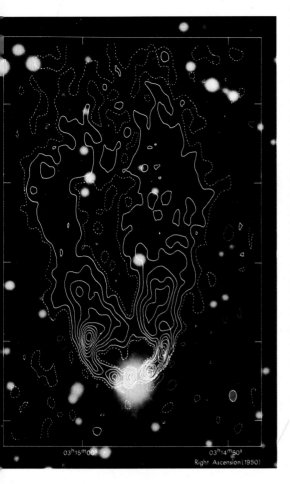

03ʰ15ᵐ00ˢ 03ʰ14ᵐ50ˢ
Right Ascension (1950)

G. K. MILEY, H. VAN DER LAAN, AND K. J. WELLINGTON, LEIDEN OBSERVATORY, THE NETHERLANDS

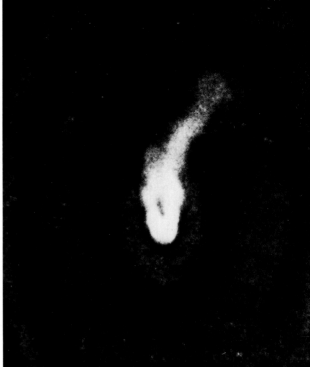

N.G.S. PHOTOGRAPHER JAMES P. BLAIR (BELOW)

"Head-Tail" Galaxies

In the control room of Mullard Observatory's Five-kilometer Telescope (a 3-mile-long array of antennas), Nobel laureate Sir Martin Ryle prepares to observe a "head-tail" galaxy. In 1968, Ryle and his colleagues noted that some galaxies give off strong radio signals from the visible areas, the heads, and weak signals that form invisible tails. A radiograph (upper right), a photograph made from radio data, shows the shape of the tail of a galaxy called NGC 1265. The tail may represent the wake of the galaxy as it moves through a sea of intergalactic gas. Such wakes may help in probing the evolution of galaxies and of the gas that surrounds them. Lines superimposed on a photograph of the galaxy (upper left) map the radio emissions that form its tail.

Bell engineers detected in their microwave receiver. Indeed, as David T. Wilkinson, one of Dicke's associates, has pointed out, that ancient radiation causes some of the flecks of "snow" we see on our television screens.

Today's universe is the cooled-down fireball, and we are, of course, inside it. In every direction astronomers' radio antennas find the whispering background from past eons. Since the radiation turns out to be remarkably uniform in all directions, astronomers conclude that the fireball was almost round, and that the universe is expanding at the same speed in all directions. This does not prove that the story of big-bang creation is true, but it provides strong support for it.

Einstein never liked the idea of creation in a single flash, and of a universe that would come to an end. He believed in a universe that was unchanging and eternal. But the equations of his General Theory of Relativity actually required a dynamic universe—it could not sit still. Since the expansion of the universe was not yet recognized when Einstein developed his theory, he deliberately introduced a term—a "cosmological constant"—in the equations to hold the universe in place. In later years he remarked that adding the term was the greatest blunder he ever made.

One of the first scientists to recognize the full significance of general relativity for an evolving universe was a young Russian mathematician, Alexander Friedmann. In 1922 he showed that Einstein's theory could explain how the universe may have expanded from a single source. Furthermore, he predicted that all galaxies should be moving away from us with speeds proportional to distance. Friedmann died of typhus in 1924, and little note was taken of his ideas for the next ten years.

In 1927 Abbé G. Lemaître, a Belgian physicist and astronomer, completely unaware of Friedmann's work, developed a similar theory. Again independently, and only a year later, the same result was derived by cosmologist H. P. Robertson in the United States. All of this theoretical work somehow escaped the attention of most astronomers. But by 1929 Edwin Hubble had concluded that the universe was expanding, and a year later, Sir Arthur Eddington came upon Lemaître's paper, which he immediately related to Hubble's discovery.

IF THE UNIVERSE is expanding from a big bang, what will its future be? The answer lies in the shape of the universe itself.

According to Einstein, the gravitational pull created by massive bodies can be expressed as a curvature of space. We know already how the mass of the sun warps the space around it—puts a dent in space—and how light rays follow this curvature to bend around the sun.

Einstein showed mathematically that in addition to such local warps and dents, the entire universe must have an all-pervading curvature. The more material present, the greater would be the curvature. And the degree of curvature is a measure of whether the universe is slowing its outward rush or growing without limit.

Throughout the 1930's Hubble struggled vainly to detect the curvature of the universe, counting galaxies to greater and greater distances. But because the curvature does not become pronounced until the distances approach the edge of the observable universe, Hubble's telescopes simply could not see far enough.

The search for the curvature effect by the continued study of the distance-velocity relation itself has been inherited by Hubble's student, Allan Sandage. From Einstein's equations, a simple connection has been derived between the red shift-distance relation and the curvature of the universe.

Space may have positive or negative curvature. A basketball is an example of positive curvature; its surface bends inward. An object with negative curvature has a surface that bends outward, like a saddle.

Some cosmologists believe the universe is "closed," and the curvature is positive. That is, it will someday stop its expansion, fall back, and then collapse. Others contend the curvature is negative, the universe is "open," and it will expand forever.

If the universe is closed, the expansion has been losing momentum much faster than if the universe were open. To measure the early slowing down, astronomers study the light from the most distant objects. Since this light left its source billions and billions of years ago, the astronomer is using his telescope to look backward in time. The Hale Telescope, which can detect as many as a million galaxies inside the bowl of the Big Dipper alone, can also see galaxies some five billion light-years away from us — or as they were about five billion years ago.

Until recently the evidence seemed to indicate that the universe is closed — that in time the expansion will slow to a halt. However, new evidence tends to support the theory of an ever-expanding universe.

If we do accept the crude evidence that the universe is closed and that the rate of expansion is slowing, the red shifts of the most distant galaxies must slowly decrease so that each year more and more galaxies come into the view of our telescopes.

An end of the expansion could come in a few tens of billions of years. For an instant, all galaxies will hang motionless, like the turning of the tide. Then, like rockets that spend their energy without escaping the pull of gravity, the galaxies will start to fall back. If astronomers could witness this dramatic change in our universe, they would see blue shifts as galaxies rush in on them from all directions. Finally, all the stars will crush inward to be cremated in a new fireball. Density and temperature will skyrocket. The fireball will become unstable, and perhaps it will explode in another big bang — and a new universe will rise like a Phoenix from the ashes of the old.

The physics of the collapse and explosion is unknown. A new universe might not resemble the old one. Physical constants, such as the speed of light, the makeup of atoms, and gravity might change. The cycle might reproduce itself, with the universe oscillating like a Yo-Yo forever.

IT TAKES BOTH an optical and a radio telescope to determine the distance to a radio galaxy. Radio telescopes alone cannot readily find the distance because radio galaxies do not emit radiation in spectral lines from which a red shift can be determined easily. So astronomers first use a radio telescope to pinpoint the galaxy's position.

Then, relying on an optical telescope, they match the galaxy with its visible counterpart. After determining the red shift of the visible source, they calculate the galaxy's distance from earth. So far, almost 50 percent of these bright radio sources have been identified with optical objects.

The discovery of Cygnus A, with a million times the radio power of our own galaxy, showed how we might search out such powerful sources at even greater distances with radio telescopes than with optical telescopes.

Suppose, for simplicity, that all radio galaxies had the same power, and that they were scattered uniformly throughout the universe. Distance and red shift would cause the more distant radio sources to appear dimmer from earth. As astronomers probed deeper and deeper into the universe for these fainter objects, they expected a certain pattern of increase in the number of radio sources.

But when Sir Martin Ryle, Britain's Astronomer Royal, looked for this phenomenon, he found even more radio sources than expected toward the edge of the universe. There appeared to be a greater concentration of radio galaxies farther away than in the vicinity of our own galaxy.

In trying to reason this out, Ryle came up with the idea that he was looking back in time to the evolution of the universe when powerful radio galaxies were more numerous than today. As he pressed further to observe still dimmer radio galaxies, he found a rather dramatic drop in broadcast signals, as though he were looking back to a time when no radio sources existed. Ryle was convinced that his observations were proof the universe is evolving — that we do not live in a steady-state universe.

These results aroused strong reactions among cosmologists. Some argued that an excess of distant radio galaxies is just an illusion created by a fortuitous scarcity of nearby radio sources. Other cosmologists contend that the galaxies are really nearby radio sources that are so weak they seem much farther away.

ANOTHER IMPORTANT CLUE to the nature of the universe could be found by determining the amount of matter it contains.

The rate at which the expansion of the universe is slowing down depends on the average density of matter in the universe. Scientists have calculated that if the average density is more than one atom in about ten cubic feet, the expansion must inevitably stop and then contract because of the pull of gravity. This average density is extremely small; it corresponds to a tiny drop of water expanding to fill a volume larger than earth. Yet the average density of matter in the universe we know is even less than this!

Astronomers have estimated the quantity of material in just the galaxies of the observable universe. This mass provides a mere two percent of the matter necessary to stop the expansion of the universe by gravitational attraction.

If the universe is indeed closed, where could the remaining 98 percent of the mass be hidden?

It could be in stars too faint to be seen on the fringes of galaxies, or in dim dwarf galaxies, or in galaxies composed of burned-out stars, or in brick-sized rocks; or it may have been swallowed up in black holes.

Many astronomers believe that the missing mass eventually will be found in clouds of gas observed outside the galaxies. But all efforts to detect this gas with optical or radio telescopes have failed.

If it is very hot, the gas may emit X rays. Astronomers have detected a diffuse X-ray glow which covers the sky and which could indicate the missing mass. If intergalactic gas is clumped into clouds, however, it can radiate X rays much more strongly than if it is evenly spread out. Much less gas would be required in cloud form to account for the X-ray background.

Kenneth Freeman of the Mount Stromlo Observatory in Australia and Gerard de Vaucouleurs of the McDonald Observatory in the Davis Mountains of west Texas may have evidence that the intergalactic gas is, indeed, clumped in clouds. They have studied "ring galaxies," which appear as a galaxy and a ring of gas side by side. They believe that such structures occur when a collision between a spiral galaxy and a cloud of gas sweeps the interstellar gas out of the galaxy like a smoke ring. The gas could contain the mass of a billion suns.

Many cosmologists disagree with the Freeman-de Vaucouleurs theory. They argue that for each spiral galaxy hundreds of such gas clouds would be required even to approach the number of random collisions between galaxies and clouds that astronomers believe they have seen.

If we cannot find the mass necessary to halt the expansion of the universe when we look at the universe as a whole, perhaps we should look at the clusters of galaxies that travel together through space. If we can examine the forces that bind these clusters, we can perhaps draw conclusions about the force that binds the entire universe together.

As clusters of galaxies cruise through space, the members of each convoy move about at random. While the Coma cluster, with more than a thousand galaxies, races away from us at 4,300 miles a second, the individual galaxies within the cluster have speeds of only a few hundred miles a second as they flit around inside the cluster. If there were no gravitational binding force, all this motion should have dispersed the galaxies in the cluster long ago. But gravity pulls on them as if they were tied to springs joined to the center of the cluster.

This tight clustering suggests that gravity keeps the individual galaxies from escaping. The strength of the gravity depends on the total mass in the cluster. Once astronomers can determine the speeds of the galaxies—which are governed by the pull of gravity—they can measure the mass. They have found that the binding mass which should exist would exceed the total mass of all the visible galaxies combined.

Other clusters show similar deficiencies in the mass of their galaxies. Where, again, is the missing mass?

Could it be in gas between the galaxies of the cluster, just as the space between stars in our galaxy is thinly filled with gas? The open spaces between galaxies in a cluster are so large that an enormous amount of gas could be there, and yet the concentration might be too small to detect. Some gas must certainly be there because it is inconceivable that all the primordial gas was used up when the galaxies condensed.

We can find graphic evidence of gas between the radio galaxies of a cluster by examining the microwave images of individual galaxies that are moving at high speeds. The pattern of radio waves shows that the magnetic field of the galaxy is swept back into an extended tail, or wake, as if the galaxy were ramming its way through a thin sea of gas. But when the density of this gas is calculated, there seems to be only a small fraction of the gas necessary to supply the gravitational "glue" to hold the galaxies in a group.

The same conclusion comes from observing X rays in the Coma cluster. The X-ray thermometer shows hundred-million degree gas, but the amount is only about two percent of the mass that would be

New galaxies from older ones: Thus theorizes Victor A. Ambartsumian (left), director of the Burakan Astrophysical Observatory in Armenia, here chatting with astrophysicist G. A. Gurzadyan. Ambartsumian suggests that new galaxies may form from explosive nuclei of older galaxies. His view differs radically from the widely held one that galaxies develop when gas clouds condense.

Most distant galaxy yet photographed, 3C 123 appears as a blur (arrows) near numbered reference stars. Eight billion light-years away, it emits more radio noise than any other known radio galaxy. What fuels this powerful energy source remains a mystery.

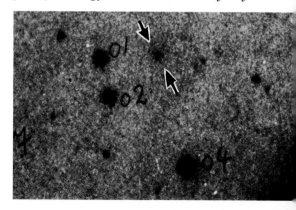

required for binding. All efforts to observe any cold or lukewarm gas have failed.

The mass cannot be hidden in a supermassive black hole because the gravitation from the hole would distort the random motions of the galaxies in a way that could be observed. If the mass were hidden in smaller black holes, there would have to be hundreds of thousands of them spread through Coma. If they are present in the Coma cluster, they should be present in our own cluster — but we have not yet found them.

When the question of gas within a cluster is examined in relationship to gas outside the cluster, the problem grows even more perplexing. The cluster must be draining gas from intergalactic space by gravitational pull. By now, the interior of the cluster should be gorged with this gas. When we fail to find much cluster gas, it must mean that there is relatively little intergalactic gas to be consumed. In effect, the amount of intergalactic gas must be a small fraction of what is required to close the universe. Is there any way to counter this argument? Perhaps a galactic wind blows out of the cluster and drives away the in-falling gas, but we have no evidence of such winds.

If we cannot find the binding mass, then perhaps the galaxies within the cluster are dispersing because of an explosion that gave birth to them — a "little big bang."

Implied in the search for missing mass is a belief that the universe is closed. Einstein thought so, and his insights still have a profound influence on our thinking. But most evidence is now pointing more and more strongly toward an open universe that will expand forever.

A basic yardstick to determine the expansion of the universe is, of course, the measurement of the red shift of an object.

The great majority of scientists accept the view that the red shifts of both galaxies and the still puzzling and powerful quasars are caused by their speeding away in an expanding universe. If the quasars are really as far away as they seem, they are producing unbelievable amounts of energy. But there are some nagging doubts. A few astronomers have persistently searched for evidence that red shifts are not completely caused by the rapid outward movement of quasars.

One of the problems in astronomy is that the brightness of an object as seen on earth depends not only on how bright the object really is, but also on how far away it is. The amount of light falls off rapidly as distance increases. A bright object far away may resemble a dimmer object closer to the observer. Thus, if the mysterious quasars are really closer than astronomers think, they would not have to produce as much power to appear as bright as they do.

ANOTHER INCREDIBLE OBSERVATION is that small compact parts of some quasars seem to be separating from each other faster than the speed of light — yet nothing should be able to exceed that speed. Some ingenious explanations for this have been put forward by Martin Rees and others. They explain the faster-than-light velocity as an illusion.

Quasar data are so contradictory that Margaret and Geoffrey Burbidge have observed, "If the normal galaxy data had turned out like quasar data, we might still not know that the universe is expanding."

Halton Arp of the Hale Observatories points out that a substantial number of quasars appear close to active galaxies. Arp contends that the probability of so many chance associations is quite small.

Arp believes that he has also identified quasars and seemingly related normal galaxies which have very different red shifts. He says that a bridge of gas seems to connect the quasar Markarian 205 with the galaxy NGC 4319. But the quasar red shift

Self-effacing genius, Albert Einstein transformed man's understanding of the universe. In 1916, he described it in terms of curved space-time. Physicists use his theory to build a model of an expanding universe.

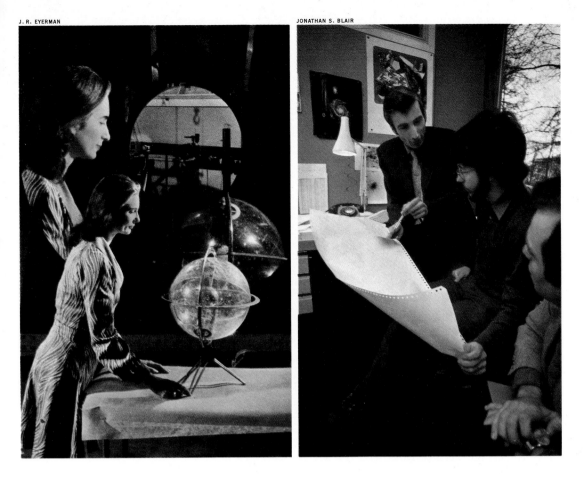

Creative thinkers probe the universe: Geoffrey and Margaret Burbidge (left) ponder the nature of quasars, a possible link in the evolution of galaxies. Beverly Lynds, reflected in a telescope mirror at the Kitt Peak optical shop, studies the dust and gas of spiral galaxies, seeking to find why our galaxy takes a spiral shape. Pointing out details on a computer readout, Martin Rees confers with students. He thinks the explosion of information about the universe may give rise to new laws of physics. Theorist Fred Hoyle (below, left) ranks as an originator and foremost proponent of the concept of a steady-state universe. Physicist Philip Morrison (below, right) believes the evidence favors the big bang.

gives a velocity of 13,000 miles per second, whereas the galaxy's velocity is only 1,050 miles per second. If the objects are truly connected, they should have the same red shift. Arp's list includes a number of such celestial "freaks," and he believes that they bring the classical interpretation of red shift into serious doubt.

John Bahcall of the Institute for Advanced Study at Princeton, one of many astronomers who disagree with Arp, has engaged him in a "Great Red Shift Debate." He dismisses Arp's peculiar examples with the suggestion that they are chance superpositions of background and foreground objects. Bahcall argues: "The skies when photographed with large telescopes reveal so many individual objects on any photographic plate that one can find almost any configuration one wants if one just hunts: even stars arranged as four-leaf clovers."

The present controversy over the distances of quasars is reminiscent of arguments about binary stars some 200 years ago. John Mitchell, a clergyman and amateur astronomer, argued that these pairs of stars must be physically related by gravity or by some special act of God. William Herschel objected. He said that the members of pairs often had greatly different brightnesses, and since he believed that all stars were of equal brightness, the members of binary pairs had to be distantly separated.

It took Herschel 36 years to change his mind. We hope it will not take us that long to solve the quasar riddle.

Perhaps it is time for astronomers to pause and wonder whether they know too much and understand too little. In 1658, Thomas Willsford said: "Thus numerous are the opinions of learned Philosophers, Geometricians, Astronomers, Geographers, Cosmographers and Navigators, and their ways so ambiguous, seldom agreeing in anything, often crossing one another, that if there be a truth in them, it is hard ... to discover what it is, and being found, difficult to follow." Many observations and theories

have fallen into place, but many phenomena still require explanation. This is the kind of situation that makes science exciting and challenging. Thus, the mysteries of the universe will continue to be pressed with every resource available to astronomers.

One new instrument has immense promise: the Large Space Telescope. Scientists expect it to be carried into earth orbit by the Space Shuttle in the early 1980's. With a large mirror (94-inch aperture), operating above distorting effects of the lower atmosphere, astronomers hope to detect galaxies a hundred times fainter than those now known. X-ray astronomers also plan to orbit a large 47-inch X-ray telescope in the mid-1980's, with power to explore clusters of galaxies possibly well beyond the range of optical telescopes.

Ground-based astronomers are not letting themselves be counted out of the action. More large telescopes at clearer sites are being equipped with supersensitive light detectors, image amplifiers, and computerized image reconstruction methods that promise to eliminate frustrating twinkling. Multiple-mirror telescopes may solve the problems of increasing mirror size by using a group of smaller mirrors in unison that equal the light-gathering power of a much larger one.

The future of astronomy is promising, indeed. Astronomers look forward to observations that will illuminate the dark areas and theories that will resolve the contradictions. The mysteries of the origin and future of the universe may never be completely solved. The search may be endless, but it will always challenge the mind and excite the imagination.

High-altitude balloon, trailing tubes used to fill it with helium, will carry aloft a gamma-ray telescope. By detecting radiation above the screening effects of the atmosphere, the instrument helps in the study of nuclear reactions in supernovae.

The Search
for Life

FOR ALL THE TREASURE of information that has come from recent exploration of the sun and the planets, we are hardly more certain of their origin and evolution than we are of the history of the universe. Half a millennium after Copernicus, we are only at the threshold of understanding the solar system and the particular place of humankind in the grand cosmic plan.

Nine planets, hundreds of known comets, and many thousands of asteroids revolve about the sun. Infinitely greater numbers of smaller bodies—meteoroids and dust particles—are held in its gravitational field. All the evidence indicates the same age for the earth, the moon, and the meteorites we have recovered: approximately 4.6 billion years. From our understanding of stellar evolution, we estimate the age of the sun itself to be about the same.

Some four billion years ago, oceans formed over vast areas of the earth, and in the next billion years or so, simple plant life evolved and an oxygen atmosphere gradually developed, primarily from plant photosynthesis. Microfossil records in sedimentary rocks date back more than two billion

Overleaf: Volcanic isle of Surtsey hurls lava into the sea off Iceland. Heat from volcanism, or ultraviolet rays, or lightning may have energized earth's early atmosphere, causing the formation of organic molecules and eventually leading to the beginnings of life in the ancient seas. HJÁLMAR R. BÁRDARSON

years, and the remains of dinosaurs 200 million years. Cro-Magnon man appeared on the scene about 35,000 B.C.

What, meanwhile, was taking place elsewhere? Do we, the descendants of Cro-Magnon man and his contemporaries, represent the only intelligence ever to contemplate the amazing universe? Or, like our sun, are we typical of an average, recurring phenomenon, in no sense unique? We may never know the answer, but in raising the question we should try to understand how the planets may have formed in our solar system, and consider how special are the conditions necessary for the support of life.

There has been no shortage of theories on planetary evolution. As long ago as 1644, René Descartes suggested that the planets formed out of a rotating disk of gas and dust. In the mid-18th century, Georges-Louis Leclerc, Comte de Buffon, proposed that they were created by the splash of a comet that ran into the sun. Somewhat later, Immanuel Kant and the French mathematician-astronomer Pierre Simon, Marquis de Laplace, independently elaborated on Descarte's view that the planets had formed from a great cloud of gas.

Around the beginning of the 20th century, geologist Thomas C. Chamberlin and astronomer Forest Moulton revived Buffon's idea of a celestial collision, but involved another star rather than a comet. They suggested that near-contact with a passing star caused great masses of solar gas to tear free. The planets condensed from this gas cloud.

Actually, stellar collisions are extremely unlikely. If the sun were represented by a golf ball, neighboring stars would be several hundred miles away, and would move only the distance of one diameter a day. Probably fewer than a hundred collisions have occurred in the Milky Way over its entire existence. Moreover, Lyman Spitzer showed that it was not possible for the escaping gas to form planets even if two stars did collide: The gas would be so hot that it would dissipate into space.

Starting with a sun already forming, Hannes Alfvén and Gustaf Arrhenius picture matter falling toward the evolving sun and gaining speed until, at some critical velocity, it becomes ionized. The now-electrified gas, trapped on the magnetic field lines of the sun, is forced to rotate with it. Grains condense out of the plasma and evolve into orbits in which they interact and eventually combine to form planets.

Despite the variety of theories, the present consensus on the evolution of the solar system favors a model in which planets formed at about the same time as the sun out of a huge cloud of gas and dust.

Details of the planetary accretion are still highly speculative. Close in, heat-resistant material presumably would survive the high temperatures to solidify into the inner planets; in the cooler outer regions, the more volatile matter would remain to form the giant planets.

If we accept the theory that the solar system was formed as part of the same process that gave birth to our sun, then it is likely that other planets have formed in the same way around other stars. We have not yet directly observed a single example, but this is not surprising. If viewed from the distance of Alpha Centauri, our nearest stellar neighbor — 4.3 light-years away — the sun would appear as a rather bright star; but earth would be not only 1.6 billion times fainter than the sun but so close to it as to be indistinguishable.

There is, however, another way to look for possible planets. In recent years, precise observations of movements of certain stars in the general neighborhood of the sun have revealed wiggly patterns that may indicate their coupling to planets. If a star has a companion body and revolves around a common center of gravity, it will appear to curve back and forth across a straight-line path in the sky. The time required for each such cycle reveals the time period of one orbit.

A few stars have been discovered that appear to have invisible partners — planets or very dim stars. The one most thoroughly studied is Barnard's Star, a red dwarf only six light-years away. Its motion across the sky in relation to the background stars is the greatest of any star known.

Peter van de Kamp of Swarthmore College has studied Barnard's Star since 1938. After years of painstaking measurements, he concluded that there is a wiggle in the star track with a period of 11.5 years, caused by a mass about equal to that of Jupiter. The width of the wiggle is comparable to the angular width of a pinhead at a distance of ten miles — testifying to the remarkable precision of van de Kamp's astrometry. Now there is some indication that a second object, about two-fifths the mass of Jupiter, is also orbiting about Barnard's Star.

The case of Barnard's Star — although it is not beyond doubt — is the most suggestive evidence we have for another planetary system in the solar neighborhood. Both Barnard's Star and the sun lie within a range that includes four stars; thus two out of four appear to have planets comparable in size to Jupiter. This is too small a sampling, of course, to make a statistical case. However, other recent studies of sunlike stars of the Milky Way do in fact suggest that about a third of them may have planets.

GIVEN THE LIKELIHOOD of other planets throughout the universe, what can we infer about their having the conditions and ingredients necessary for life?

In earlier times, people believed life could arise spontaneously. Anaximander, a Greek philosopher of the sixth century B.C., taught that life arose from undersea mud and men emerged from fish. In the 17th century the Belgian scientist Johannes Baptista van Helmont soberly explained that mice could be produced by a process involving grain and dirty laundry: "And what is more remarkable, the mice ... are neither weanlings nor sucklings nor premature, but they jump out fully formed."

Only in the middle of the last century

did Pasteur's proof of the existence of airborne "germs" and the impossibility of growth of microorganisms in sterile media put an end to theories of spontaneous generation. Early in this century the Swedish chemist Svante Arrhenius proposed the concept known as "panspermia"—that terrestrial life did not originate here but drifted to earth from remote space. He believed that spores of living organisms moved from world to world, propelled by radiation pressure. But it is difficult to conceive of the wandering spores' surviving all the hazards of space—the lethal effects of ultraviolet, X-ray, and cosmic-ray bombardment over long periods of time.

More recently Tommy Gold has whimsically suggested another means of transporting microorganisms: Suppose that representatives of advanced civilizations have visited uninhabited planets; they might well have left their contamination behind. Gold's "garbage theory" pictures the visitors picnicking on a virgin planet and leaving behind their crumbs to start a new cycle of life.

FROM PASTEUR'S DAY until well into the 20th century, little scientific progress was made into the question of the origin of life. Then in the 1920's the Russian biochemist A. I. Oparin and the British biologist J. B. S. Haldane independently suggested a chemical evolution of life that began in the primitive earth's atmosphere. They reasoned that a source of energy such as ultraviolet light from the sun could have created organic compounds which dissolved in the oceans to form a "hot dilute soup." These compounds would then combine into more complex molecules and eventually evolve into living organisms.

Contemporary studies of the origin of life have been stimulated by the classic research of Stanley Miller, while a student of chemist Harold Urey at the University of Chicago. In 1953, starting with a heated mixture of hydrogen, methane, ammonia, and water vapor to simulate earth's early atmosphere, Miller determined the organic compounds produced by the passage of a 60,000-volt lightning-like spark through the gases. Several kinds of organic molecules were formed, including the amino acids glycine and alanine.

In recent years, scientists in many laboratories have conducted similar experiments that have given strong support for a chemical origin for life on earth.

It is important to note that these experiments are conducted without oxygen, which it is believed was not present in the primitive earth's atmosphere.

Additional evidence of chemical evolution comes from the observation of molecules in interstellar space. Simple molecules were discovered years ago by spectrographs attached to ground-based telescopes. But radio astronomy has lately revealed a host of molecules ranging from water and ammonia to such organic compounds as formaldehyde, cyanoacetylene, and ethyl alcohol. Large molecules in interstellar space were once unsuspected, for the gas there is so thin that the rate of occurrence of collisions in which atoms stick together to form molecules is extremely slow. Apparently dust particles trap the atoms and catalyze their combination into molecules. Each year now brings new additions to the list.

What role may these primitive molecules play in the evolution of life? Undoubtedly most of them would be destroyed by the heat accompanying the coalescence of an earthlike planet; yet some might be preserved in such natural refrigerators as comets. Fred Whipple's idea that comets are "dirty snowballs"—mostly a mix of ordinary ice and dust—has been confirmed by recent studies of comets, including Kohoutek. Several prebiological molecules have been detected in comets, and amino acids have been found in a type of meteorite called carbonaceous chondrite.

After a planet has had time to develop its life-supporting environment, these "seeds of life" may rain down from space and survive. Fred Hoyle suggests that we may owe our existence to the snowballs, although most scientists believe their contribution — if any — to life on earth would be very small.

Dr. Leslie E. Orgel of the Salk Institute for Biological Studies thinks it more likely that life evolved on the earth than that it reached us from space. "However," he adds, "the latter point of view is an interesting one and should be entertained by students of the origins of life, at least on sleepless nights."

The observation of organic molecules in space, and the laboratory production of organic material under conditions that simulate the primitive earth environment, suggest that the same chemical processes occur in many places — and therefore that complex prebiological material might be widespread throughout the universe. Life elsewhere, if it exists, is probably chemically similar. Professor George Wald tells his students: "Learn your biochemistry here and you will be able to pass examinations on Arcturus."

IN THE FOURTH CENTURY B.C., the Greek philosopher Metrodoros of Chios declared: "To consider the earth as the only populated world in infinite space is as absurd as to assert that on a vast plain only one stalk of grain will grow." This view is echoed by Richard Berendzen of American University, who says "the question has become not so much one of *if* as of *where*."

To support life similar to that on earth, a planet must have liquid water and a suitable atmosphere. The prevailing temperature ideally should be above the freezing point of water and below its boiling point, and should not vary too greatly. Perhaps one percent of all stars have planets that can support life. For the Milky Way, this means about two billion suitable planets — and very likely one within 20 light-years of our sun.

Closer to home, evidence concerning questions of extraterrestrial life has been obtained by spacecraft dispatched to the moon, Mercury, Venus, Mars, and Jupiter.

The moon, as Apollo's television viewers know well, has no atmosphere and receives the full blast of solar radiation from X rays to infrared, as well as the bombarding particles of the solar wind. No trace of life there has been found in samples returned by the lunar explorers.

Conditions on Mercury are even more severe. And on Venus, the average surface temperature is nearly 500° C. Jupiter, Saturn, Uranus, and Neptune are very different physically and chemically from earth. We have little basis yet for serious consideration of life on these planets, although scientists do speculate that some form of life might be able to exist in Jupiter's atmosphere.

The first real search for life beyond earth is the Viking mission. In the summer of 1976 it will land equipment on the surface of Mars to measure surface and atmospheric composition and survey the vicinity for evidence of organisms based on carbon chemistry — as is life on earth. If evidence of living material is found on Mars, we can reasonably conclude that the development of life on earthlike planets elsewhere in the universe may well be commonplace rather than a rare accident. And we can then proceed with the search for intelligent communication with reinforced enthusiasm.

DISCOVERY of any form of life would be an exciting breakthrough, but it is the idea of contact with advanced civilizations that challenges our imaginations and endlessly stimulates our curiosity.

Although travel to the vicinity of another populated planet would take many thousands of years, our present radio technology can send detectable signals over the distances to the nearest stars and expect an answer — if there is someone there — well within a human lifetime.

In April and June of 1960, Frank Drake made an effort to search for galactic radio signals of intelligent origin. Project Ozma, as it was called *(Continued on page 192)*

PAINTING BY ARTHUR HILL, TRW SYSTEMS (ABOVE); AMES RESEARCH CENTER, NASA
(BELOW AND OPPOSITE); NATIONAL RADIO ASTRONOMY OBSERVATORY

Eyes and Ears on the Universe

Some of the most sensitive and gigantic devices ever conceived may soon explore the distant reaches of space.

The unmanned satellite at top will relay information on such high-energy emissions as gamma rays and X rays, which some celestial objects send out continuously and others emit in violent bursts. The first of three High Energy Astronomy Observatories, it will rocket into orbit in 1977.

Scientists will zero in on the most distant objects, and produce radio pictures, by com-

bining the signals received by 27 antennas on a Y-shaped rail track (opposite, bottom). Now under construction near Socorro, New Mexico, the facility bears the descriptive name Very Large Array.

To listen for evidence of other civilizations, scientists propose a circular cluster of radio telescopes—initially a few, eventually perhaps hundreds covering a miles-wide area (opposite, top). The system, called Cyclops, would employ steerable antennas (above), each larger in area than a football field.

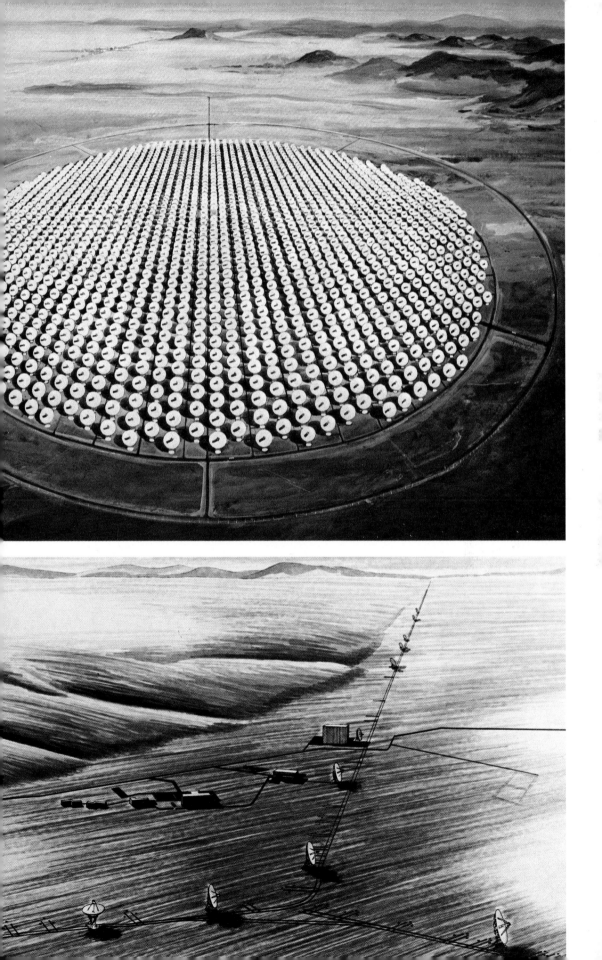

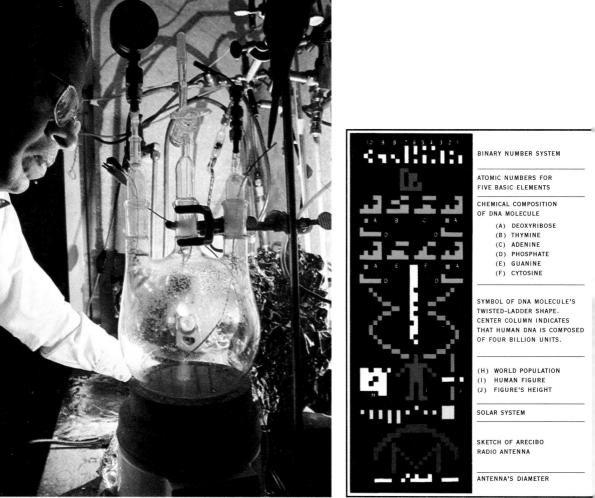

BINARY NUMBER SYSTEM

ATOMIC NUMBERS FOR
FIVE BASIC ELEMENTS

CHEMICAL COMPOSITION
OF DNA MOLECULE

    (A) DEOXYRIBOSE
    (B) THYMINE
    (C) ADENINE
    (D) PHOSPHATE
    (E) GUANINE
    (F) CYTOSINE

SYMBOL OF DNA MOLECULE'S
TWISTED-LADDER SHAPE.
CENTER COLUMN INDICATES
THAT HUMAN DNA IS COMPOSED
OF FOUR BILLION UNITS.

(H) WORLD POPULATION
(I) HUMAN FIGURE
(J) FIGURE'S HEIGHT

SOLAR SYSTEM

SKETCH OF ARECIBO
RADIO ANTENNA

ANTENNA'S DIAMETER

N.G.S. PHOTOGRAPHER OTIS IMBODEN (ABOVE); NATIONAL ASTRONOMY AND IONOSPHERE CENTER
(ABOVE, RIGHT); ASTRONOMY MAGAZINE PAINTING BY ADOLF SCHALLER © 1975 ASTROMEDIA CORP.

Life's building blocks brew in a flask of ammonia, methane, and water vapor, thought by University of Maryland chemist Cyril Ponnamperuma to simulate earth's early atmosphere. When sparked by an electrical discharge, the gases form the basis of nucleic acids and proteins. Scientists believe that more than three billion years ago such organic compounds evolved into simple organisms.

The first intentional radio space message (above, right), beamed in November 1974 from the Arecibo Observatory, announces man's existence to any intelligent life capable of receiving and interpreting the code. After extensive improvements at Arecibo—it now sends the strongest signal leaving earth—astronomers broadcast basic information about our planet. The message raced past Pluto and sped out of our solar system within 5 hours and 20 minutes.

Star-spangled blackness envelops the meteoroid-pocked remains of Pioneer 10—the first of two spacecraft to inspect Jupiter. In an artist's conception of the vehicle as it might appear millions of years from now, Pioneer cruises through uncharted oceans of space. The gravitation of a star or a planet may eventually capture the craft. It carries a plaque engraved with coded information about its origin, and a sketch of a man and a woman.

(after the Princess of Oz in L. Frank Baum's stories) was conducted with an 85-foot telescope at Green Bank aimed at two of the nearest stars. The results were negative; but of course two stars hardly constitute a search —more like several million must be tried before success can be expected.

At the time of Ozma, it took great courage to devote 200 hours to what many would call a crazy idea. But in 15 years our perspectives have changed, and there is among scientists and the public as well a considerable enthusiasm for a serious new effort. Perhaps the universe is filled with voices calling from star to star.

Drake is now director of the largest radio telescope in the world, the 1,000-foot dish at Arecibo, Puerto Rico. In November 1974 the newly improved telescope beamed into space a coded message that could be detected by a similar telescope almost anywhere in the galaxy.

If we take a long-range view, is there a better way of searching for the needle in the celestial haystack?

Ronald Bracewell of Stanford University has proposed a strategy he suspects a more advanced civilization may already be trying: dispatching automated spacecraft to thousands of the most likely stars. Within a few centuries we may be able to send such vehicles to stars as far away as a hundred light-years. Travel time might take up several thousand years; but once in orbit, the probe would draw its power from the star's light. Automatically transmitted radio messages to the star's planets would be loud and clear, and could hardly fail to be recognized by intelligent monitors.

Meanwhile, Bracewell urges that we be alert to all cosmic radio signals for evidence of such probes from distant civilizations.

Bernard Oliver, a vice president of the Hewlett-Packard Company, has suggested building a giant array of radio telescopes to search for extraterrestrial intelligent signals. Called Cyclops, the project plan was the result of a NASA study he directed in 1971.

Starting with a few antennas, Cyclops would examine likely nearby stars for signals from their planets. Year by year, as the array grew, more distant stars could be searched. Success might be achieved early, or the search might take decades and require a thousand antennas or more.

Discovery of an extraterrestrial intelligence would be an event of transcending importance in human history. Most likely the exchange would take hundreds of years, and the first contacts would simply establish a language of communication and transmit the most elementary information. As with most proposals for new human undertakings, there are pessimistic and optimistic attitudes. Some people think the cultural shock of discovering a much more advanced civilization might be psychologically damaging to ours. But the optimistic view is that we could gain a new sense of purpose in life and a strengthened faith in the future.

———————————

This world was once a fluid haze of light,
Till toward the centre set the starry tides,
And eddied into suns, that wheeling cast
The planets: then the monster, then the man.

Lord Tennyson's lines sum up our story of the amazing universe. Man has been part of it all for a brief instant of cosmic time. His observations are fragmentary and often ambiguous. Yet from these fugitive perceptions he has pieced together a remarkable picture of the universe and the physical laws that govern it.

Albert Einstein once remarked that the most incomprehensible thing about the universe is that it is comprehensible!

Newborn star shines beyond a swirling cloud of organic and inorganic molecules (foreground) coalescing into a planet. Astronomers have detected such molecules in many areas of the galaxy, leading them to speculate that the same chemical processes that led to life on earth may have occurred elsewhere.

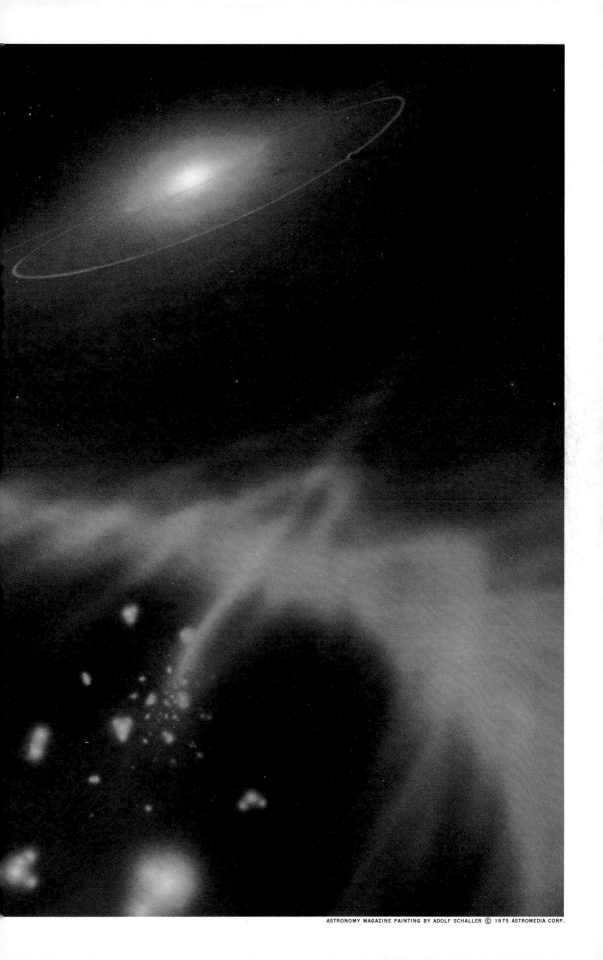

Burro-power recovers a gamma-ray telescope launched by balloon to search for emissions from the Crab Nebula. The instrument rose 125,000 feet above New Mexico, then landed among remote, rugged mountains.

Glossary

Binary star: one of a pair of stars orbiting each other under mutual gravitation.

Black hole: volume of space, only a few miles in diameter, from which no matter or radiation—including light—can escape because of the intense gravity of an exceedingly dense collapsed star at its center.

Cepheid variable: supergiant star whose light fluctuates, from dim to bright and back again, in a predictable time period.

Cosmic rays: highly energized atomic particles that travel through the galaxy at speeds approaching that of light.

Electromagnetic spectrum: complete range of wavelengths and frequencies caused by varying electric and magnetic fields: from short gamma rays through X rays, ultraviolet, visible light, and infrared to long radio waves.

Electron: elementary particle with a negative electrical charge, normally orbiting an atomic nucleus.

Escape velocity: minimum speed at which an object must move to escape the gravitational hold of another object.

Fusion: nuclear process that builds up heavier elements from lighter ones by the fusing of atoms, thus releasing great amounts of energy.

Galaxy: system of billions to hundreds of billions of stars, along with nebulae and other interstellar matter, held together by gravitational attraction.

Giant: large star of high luminosity, tens to hundreds of times the size of the sun.

Globular cluster: large, spherical group of stars within a galaxy.

Gravitation: force of attraction that matter exerts on other matter.

Ionization: process by which an atom loses or gains one or more electrons and becomes electrically charged.

Light-year: distance light travels in one year at a rate of 186,282 miles a second—about six trillion miles.

Luminosity: brightness of a celestial object.

Magnetic field: space around an object in which its magnetic force can be detected. The earth's magnetic field funnels charged particles from the sun into the polar regions to produce colorful aurorae, the northern and southern lights.

Magnitude: number indicating the brightness of a star relative to other stars.

Mass: total amount of matter in an object.

Milky Way: home galaxy of our solar system; also, the band of light from stars and nebulae that arches across the sky.

Nebula: cloud of interstellar gas and dust.

Neutrino: particle of no mass or charge that carries away energy from certain nuclear reactions and travels at the speed of light.

Neutron: elementary particle with no electrical charge, present in all atomic nuclei except that of hydrogen.

Neutron star: extremely dense collapsed core of a star that exploded; it consists almost entirely of neutrons.

Nova: star that violently erupts, briefly increasing its brightness by hundreds to millions of times.

Photon: smallest unit of light or other electromagnetic energy.

Polarization: aligning of the vibrations of light to one plane.

Proton: elementary particle with a positive electrical charge; the nucleus of hydrogen, and present in all other atomic nuclei.

Protostar: concentration of interstellar gas and dust condensing to form a star.

Pulsar: source of radio signals (and in some cases, optical or X-ray emission) that pulsates in a rapid, precise sequence; generally accepted to be a spinning neutron star.

Quasar: extragalactic starlike source of great radio power and extreme optical brightness.

Radio galaxy: galaxy that emits its predominant energy in radio waves—hundreds to millions of times as much as the Milky Way.

Relativity: concept of Albert Einstein that is used as a basis for theoretical models of the universe. His Special Theory, published in 1905, holds that motion, time, and distance are not absolute but relative to moving frames of reference. His second or General Theory, published in 1916, extends the Special Theory to include acceleration and gravitation.

Spectrum: array of colors obtained when light is dispersed by a prism or diffraction grating into its component wavelengths.

Supergiant: large star of the highest luminosity class known.

Supernova: massive star that explodes, releasing hundreds to millions of times more energy than a nova eruption.

Synchrotron radiation: light and other radiation produced by electrons spiraling at extreme velocity through a magnetic field.

Telescope resolution: degree to which details of an image are made distinguishable.

White dwarf: compact core of a star like the sun as it nears the end of its life cycle.

X-ray star: unusual star whose predominant emission is in the form of X rays.

Index

Boldface indicates illustrations

Library of Congress ⌖ Data

Friedman, Herbert, 1916-
    The Amazing Universe.

    Bibliography: p. 199
    Includes index.
        1. Astronomy — Popular works.
    2. Astronomy — Pictorial works.
    I. National Geographic Society,
Washington, D. C. Special Publications
Division.   II. Title.
QB44.2.F77   520   74-28806
ISBN 0-87044-179-5

Acknowledgments

The Special Publications Division is grateful to the individuals, organizations, and agencies named or quoted in the text and to those cited here for their generous cooperation and assistance during the preparation of this book: Helmut A. Abt, Richard Berendzen, Von Del Chamberlain, Frank D. Drake, Edward Fomalont, W. Kent Ford, Jr., Sidney W. Fox, Owen Gingerich, Theodore R. Gull, Peter D. Jackson, Yoji Kondo, Frank J. Low, William C. Miller, David L. Moore, N. Paul Patterson, Cyril Ponnamperuma, Allan Sandage, Maarten Schmidt, Neil R. Sheeley, Jr., Bradford A. Smith, Barry E. Turner, Gart Westerhout, Fred L. Whipple, Richard S. Young.

Additional Reading

George O. Abell, *Exploration of the Universe;* Isaac Asimov, *Asimov on Astronomy;* Bart J. and Priscilla F. Bok, *The Milky Way;* Ben Bova, *The New Astronomies;* Nigel Calder, *Violent Universe;* George Gamow, *A Star Called the Sun;* Robert Jastrow and Malcolm H. Thompson, *Astronomy: Fundamentals and Frontiers;* I. M. Levitt, *Beyond the Known Universe;* Donald H. Menzel, *Astronomy;* Cyril Ponnamperuma, *The Origins of Life;* Cyril Ponnamperuma and A. G. W. Cameron, *Interstellar Communication;* Carl Sagan, *The Cosmic Connection;* Harlow Shapley, *Galaxies;* I. S. Shklovskii (Josef Shklovsky) and Carl Sagan, *Intelligent Life in the Universe;* Otto Struve and Velta Zebergs, *Astronomy of the 20th Century;* Charles A. Whitney, *The Discovery of Our Galaxy.* N.G.S. Special Publication: William R. Shelton, *Man's Conquest of Space.* In NATIONAL GEOGRAPHIC: Thomas Y. Canby, "Skylab, Outpost on the Frontier of Space," October 1974; Edward G. Gibson, "The Sun As Never Seen Before," October 1974; Donald H. Menzel and Jay M. Pasachoff, "Solar Eclipse," August 1970; Kenneth F. Weaver, "Mystery Shrouds the Biggest Planet," February 1975, "The Incredible Universe," May 1974, "Journey to Mars," February 1973, and "Voyage to the Planets," August 1970. Readers may also want to consult the National Geographic Index.

Composition for *The Amazing Universe* by National Geographic's Phototypographic Division, Carl M. Shrader, Chief; Lawrence F. Ludwig, Assistant Chief. Printed and bound by Fawcett Printing Corp., Rockville, Md. Color separations by Colorgraphics, Inc., Beltsville Md.; Graphic Color Plate, Inc., Stamford, Conn.; The Lanman Co., Washington, D. C.; Progressive Color Corp., Rockville, Md.; J. Wm. Reed Co., Alexandria, Va.